Teaching Science in Diverse Settings

Studies in the Postmodern Theory of Education

Joe L. Kincheloe and Shirley R. Steinberg
General Editors

Vol. 150

PETER LANG
New York • Washington, D.C./Baltimore • Bern
Frankfurt am Main • Berlin • Brussels • Vienna • Oxford

Teaching Science in Diverse Settings

Marginalized Discourses and Classroom Practice

EDITED BY
Angela Calabrese Barton
& Margery D. Osborne

PETER LANG
New York • Washington, D.C./Baltimore • Bern
Frankfurt am Main • Berlin • Brussels • Vienna • Oxford

Library of Congress Cataloging-in-Publication Data

Teaching science in diverse settings: marginalized discourses and classroom practice / edited by Angela Calabrese Barton and Margery D. Osborne.
p. cm. — (Counterpoints; vol. 150)
1. Science—Study and teaching. 2. Critical pedagogy. 3. Postmodernism and education. I. Barton, Angela Calabrese. II. Osborne, Margery D. III. Counterpoints (New York, N.Y.); vol. 150.
LB1585 .T44   507—dc21   00-036393
ISBN 0-8204-4990-3
ISSN 1058-1634

Die Deutsche Bibliothek-CIP-Einheitsaufnahme

Teaching science in diverse settings: marginalized discourses and classroom practice /
ed. by: Angela Calabrese Barton and Margery D. Osborne.
−New York; Washington, D.C./Baltimore; Bern;
Frankfurt am Main; Berlin; Brussels; Vienna; Oxford: Lang.
(Counterpoints; Vol. 150)
ISBN 0-8204-4990-3

Cover design by Joni Holst

© 2001 Peter Lang Publishing, Inc., New York

All rights reserved.
Reprint or reproduction, even partially, in all forms such as microfilm, xerography, microfiche, microcard, and offset strictly prohibited.

Contents

List of Illustrations   vii

List of Tables   ix

Foreword: Toward a Political Philosophy of Science Education
      William C. Kyle, Jr.   xi

Introduction
      Angela Calabrese Barton and Margery D. Osborne   1

1   Marginalized Discourses and Pedagogies: Constructively Confronting Science for All in Classroom Practice
      Angela Calabrese Barton and Margery D. Osborne   7

PART ONE   ELEMENTARY SCIENCE EDUCATION AND SCIENCE TEACHER EDUCATION   **33**

2   The Magical and the Real in Science and in Teaching: Joy and the Paradox of Control
      Margery D. Osborne and David J. Brady   35

3   A Room of One's Own: Concrete and Conceptual Spaces
      Heidi Bulmahn Barker   59

4   Teaching in the Interface: Reflections on Urban Science Teaching
      Kathleen St. Louis, Tanahia Burkett, and Angela Calabrese Barton   79

Contents

5 A Paralogical Affirmation of Emotion's Discourse
   in Science Teaching
   Michalinos Zembylas   99

6 Visions of "Science for All" in the Elementary Classroom
   Elaine Howes   129

**PART TWO   SECONDARY SCIENCE EDUCATION
AND SCIENCE TEACHER EDUCATION   159**

7 Re/Writing Science from the Margins
   Gaell M. Hildebrand   161

8 Rethinking Science and Assessment
   Dana Fusco   201

9 Guinea Pig Pedagogy: Critiquing and Re-embodying
   Science/Education from Other Standpoints
   Matthew Weinstein   229

10 Promoting Inclusive Science Education through
   Professional Development: Challenges Faced in
   Transforming Content and Pedagogy
   Julie A. Bianchini and Lynette M. Cavazos   259

11 Feminisms, Sacred Stories, and Multiple Voices
   Sharon Parsons   295

12 Sociocultural Constructivism, Courage, and
   the Researcher's Gaze: Redefining Our Roles as
   Cultural Warriors for Social Change
   Alberto J. Rodriguez   325

List of Contributors   351

Index   355

Illustrations

2.1 Jame's fox   40
2.2 Scott holding his skateboard   42
2.3 Darius's truck   43
2.4 Thialand photograph   46
2.5 Two girls playing with cornmeal   47
2.6 Boy playing with cornmeal   48
2.7 Girl playing with cornmeal   49
2.8 Shape shifters   52

Tables

1.1 Comparative Summary of Critical Science Education   21
7.1 Writing Tasks Used by Collaborating Teachers   169
8.1 Objectives of Science from Several Perspectives   217

Foreword

Toward a Political Philosophy of Science Education

William C. Kyle, Jr.

This edited volume by Angela Calabrese Barton and Margery Osborne appears at an important time for science education. Throughout our global science education community, the rhetoric of 'science education for all' is juxtaposed to the reality of 'science education for the privileged'. It is imperative for science educators to begin to reflect upon: the history of the discipline of science education and the ways in which it has been enacted in the process of schooling; the hegemony of logical positivism and technical rationality and its uncritical acceptance within the discipline of science education; the ways in which science and science education have marginalized both learners and citizens in society; the ways in which traditional interpretations of knowledge have been connected to power and the relationship between power and knowledge; and, finally, the ways in which the discipline of science education can be transformed such that the political and social aspects of science are moved to the center of philosophical reflection upon the sciences, from its present position somewhere beyond the margins. Science educators must articulate a vision of and assume agency for an education in science that creates opportunities for self- and social-empowerment, whereby all learners are able to engage in participatory action-taking in a democratic society (see Kyle 1991, 1999a, 1999b).

This book offers a stark and refreshing contrast to the neoconservative agenda that permeates present-day international educational policy making, in which the notion exists that science education will be transformed by implementing standards, assessing students' technical knowledge and understandings, and engaging in cross-national comparisons of student achievement. Such an agenda is theoretically impoverished and politically visionless from three perspectives. First, this agenda ignores the wider complex of socio-cultural and political factors that influence the ways schooling is structured to benefit some students at the expense of others. Second, it neglects the fundamental issues of the place of science in any larger social context and it fails to acknowledge the political situatedness of science. Finally, it invokes a narrow image of science by valuing the technical interests of the empirical-analytical sciences at the expense of the practical interests of the hermeneutic-interpretive sciences and the emancipatory interests of the critical sciences. These issues are at the very heart of the values and visions expressed by the authors of the chapters herein.

Historically, notions of curriculum seek to appropriate children—and adults—to take part in life as it is today. This flawed orientation ignores the reality of rapid social and cultural change. Further, this orientation perpetuates dominating ideological discourses, thereby enabling educational systems to function as a political means to maintain and appropriate knowledge and power. And, while science educators have been engaged in the process of curriculum reform for the past several decades, for the most part they have engaged in mainstream reform. Mainstream educational discourse is "constructed within the parameters of our dominant social, economic, cultural, and political arrangements, including the limits of certain research traditions" (Stanley, 1992, p. 2). Alternatively, the discourse of critical pedagogy challenges the basic assumptions of mainstream reform initiatives, draws upon other research traditions, and poses an alternative reform agenda (e.g., Apple, 1985, 1986, 1990, 1992; Freire, 1985; Giroux, 1988a, 1988b, 1992; Giroux & McLaren, 1989; Grumet, 1988; Pinar, 1988; Pinar & Reynolds, 1992). I believe educators have the fundamental obligation to explore divergent ideas, including those that are radical. By examining marginalized discourses, the authors herein offer inspiration to individuals wishing to break away from mainstream education discourse.

Critical pedagogy includes, but is not restricted to, the "new sociology" of education, reconceptualist curriculum theory, cultural studies, feminist theories, critical theory, and various forms of postmodern and poststructuralist analysis. This scholarship, while far from monolithic, represents a significant knowledge base to orient the reform of education; yet, this discourse tends to remain on the periphery of the debate over educational reform. Why? How can we ensure that the questions of human empowerment and social justice are addressed among educational theorists so that the theoretical and ethical debates within curriculum theory will be advanced? Are there persistent goals, issues, and problems that science educators ought to be debating in the context of current calls for reform? I believe so. Science educators ought to investigate the relationship between contemporary forms of critical pedagogy and social reconstructionism, as they relate and contribute to the construction of a radical theory of education. Central to the reform of science education should be an articulated language of possibility and the critical competence necessary to reveal and deconstruct forms of oppression.

Science educators ought to ensure that their efforts are ultimately oriented toward self- and social-empowerment. This would contrast with the technical, rational, objectivist perspective that has dominated science education reform and science education research to date. With this radical turn, science educators would need to "raise important questions regarding the relationship among knowledge and power, learning and possibility, social criticism and human dignity, and how these can be understood in relation to rather than in isolation from those practices of domination, privilege, and resistance" (Giroux, 1992, p. 242), which are at work in the context of students' daily lives.

I propose that science educators move toward a political philosophy of science education in the spirit invoked by: Maxine Greene when she states that "philosophical critique takes us into the examination of ideologies and their coercive effects on thinking" (1995, p. 60); Paulo Freire, who—when concerned with the ways in which certain situations are perceived as hopelessly limiting and obstructive—talks of the need to move from abstract formulations to concrete renderings of the situation in such ways that "individuals begin to behave differently with regard to objective reality, once that reality has ceased to look like a blind alley and has taken on its true aspect: a challenge which human beings must meet" (1983, p. 86–87); and John Dewey, who sought a philosophy starting afresh from the most active tendencies of its day, thereby giving a large place to the "practical and personal", and who viewed knowledge as making a difference and knowing as a "change in reality" (1931, p.54). Each of these individuals viewed social critique as an ongoing effort to overcome false consciousness by rejecting an absolute and static view of reality and its resulting subject-object separatism.

In 1990, in the midst of a conversation with Paulo Freire regarding issues of school reform and the process of schooling, he offered the assertion that "Changing reality is very difficult, but it is possible." His statement resonated well with my personal experiences working with school districts on science reform projects. An essential aspect of my work in school-based science reform projects is facilitating the process of challenging ideological perspectives and changing reality. Changing the reality associated with such fundamental questions as "What is science?" What is education?" "What are the central tenets of an education in science?" and "What are the power relationships associated with the process of schooling and access to knowledge?" In the course of the conversation, Freire offered insight into what he envisioned was necessary for facilitating the process of changing reality: "There must be a formation—a coming together—of educators. We must know the political reasons for why we are doing what we are doing." Again, the collection of articles in this edited volume engage readers in the process of reflecting upon the political reasons for why we do what we are doing.

Historically, science education as a discipline has been dominated by an uncritical acceptance of logical positivism. Intriguing as it may be, just when the practice of science was becoming less influenced by logical empiricism, science education emerged as a discipline. A review of the literature in science education through the early-1980s reveals that science educators perceived of themselves as being more aligned with 'science' than with 'education'. As a result, there was a failure to acknowledge the significant differences in style, subject matter, rhetoric, and especially results between the natural sciences and the human studies (see Habermas, 1972). By adopting an image of science that was more a part of history than present day practice, early science educators failed to recognize how post-empiricist philosophy of science implicitly reveals just how closely traditional interpretations of knowledge were connected to an understanding of power and of the relation between power and knowledge. Scientific practices are political in ways that are central to their epistemic success. Thus, students and citizens alike have been denied access to this essential feataure of science; they have been led to reconstruct the development of science as a steady accumulation of results with the evidence to support them. In general, students have been denied the opportunity to experience the conceptual reformulations and the shifts in interests and standards that have marked the actual history of science. The social and political process of science has effectively been obliterated from the science curriculum by reformulating science into a single, collective, consistent account of the 'progression of science'. Ødegaard and Kyle (2000) note that:

> When students experience an education in science under the guiding interests of instrumental reason, they are denied a form of knowledge that enables them to

develop and change culture, including the culture of science. It is imperative that students have the opportunity to experience the hermeneutic and critical sciences in the context of their formal education in science. The goal of science education should be to facilitate both students' and the public's ability to identify possibilities, to seek challenges, to use their imagination, to transform. Students' experiences with science ought to be self-involving, socially just, and emancipatory.(p. 23)

What the discipline of science education lacked—and continues to lack—is a political philosophy of science education. I would contend that this is the primary reason why reforms in science education have a history of failure. Even reform initiatives that attempted to integrate social and political aspects of science, such as the 'science-technology-society movement', ultimately failed to have widespread acceptance amongst either science educators or practitioners. By questioning the foundational canons of science and science education, the authors open the door to engaging in discourse related to re/shaping the discipline of science education.

The totality of an education in science is equally as much oriented toward social justice, critical democracy, empowerment, action-taking, and investing in our future's intellectual capacity as it is about constructing conceptual understandings of the world. The essence of a learner's education in science must be linked to an understanding of the nature of science and learners must be afforded the opportunity to experience the meaning of "education" in science education. I assert that we must transform how we see science education so that it is congruent with both how we see science and how we see education.

Science is about knowing. However, all knowing (therefore all conceptual thought—all theory, concept, and belief) leads to contradictions. One of the most profound problems with "Reality" is that it is changing—constantly. In essence, scientists' search for knowledge is changing constantly as well. The search for knowledge—understanding the world and the universe—is dynamic! In the context of this search for knowledge, scientists believe in an external world, simply because it is the task of scientists to observe the world "out there" so that conclusions about phenomenon may be drawn. Thus, the process of science is not a quest for absolute Truth or Reality; scientific knowledge is tentative and subject to change; and the practice of science rests upon a number of taken-for-granted metaphysical presuppositions about the external world. How we see science is essential to how we construct our images of science education.

Education is about hope, dreams, aspirations, and struggle. Hope for a better life, dreams of what the world might be like, aspirations for future successes, and struggle over how to understand and overcome obstacles while searching for and achieving that better life. Education is about expanding upon the knowledge of life world experiences that learners bring to situations; it is about opening doors, opening minds, and opening possibilities. Education must be *for* something. But what? Education ought to be for the purpose of fostering critical and participatory democracy, enabling students to recognize that the world that is being presented to them is in fact a world that is being made—it is changing constantly—thus, for this very reason, it can be changed, it can be transformed, and it can be reinvented. Education frees the mind to create the future. How we see education is equally as important to how we construct our images of science education.

The constructed images of science education ought to be just that—images—not a singular, universal image. If education is about expanding upon the knowledge of life world experiences that learners bring to situations, then a universal image of science education is not possible. Science education must be contextualized and must be linked to the life world experiences of learners. These multiple images of science education ought to be open to multiple forms of research and knowledge production and take issues of locality, cultural values, and social justice seriously. Harding (1996), writing about a transformed science, uses the term "borderland epistemology" to signify the valuing of different understandings of the world that diverse cultures produce. A transformed science education would be personally relevant to people living in naturalistic environments; would acknowledge the political, social, and cultural forces at work in scientific labor; and would be linked to local contexts and issues of sustainability. The notion of locality and the social construction of science can facilitate creating a far more reflective and insightful science education.

Modernity has affected all aspects of Western life, all institutions, including education. The perception of education as a vehicle for modern development is based on the idea that schooling will affect people's beliefs, abilities, and values; thus, helping children develop abilities that will prepare them to function in a modern society (Carnoy & Samoff, 1990). Accordingly, individuals are encouraged to reject their backgrounds and traditional privileges and to accept a new "elite recruited through the impersonal ordeal of competitive examinations" (Touraine, 1995, p. 12). In essence, schooling in the post-Enlightenment era has emphasized learning of that which is already defined as knowledge, not the production of knowledge. In fact, the educational system has become the crucial aspect of the reproduction of technical rationality. Regrettably, during this period of time, education has emerged as a new invisible commodity. Perhaps "the most decisive reproductive consequence of educational systems derives from the complicity in the perpetuation of a positivistic scientific worldview that systematically blocs the formation of critical consciousness, even as it legitimates itself in liberal democratic and pluralistic terms" (Morrow & Torres, 1995, p. 237). An overt recognition of the social influence of the project of modernity provides an excellent vantage point to better appreciate the complex ways that power shapes knowledge. Similarly, if we shift our focus to science education, an overt recognition of the social influence of the project of modernity should enable educators to see and deconstruct the educational process in a way that has the potential to lead to socially just and culturally sensitive practices.

Transformative possibilities may emerge when individuals construct an understanding of the processes associated with how knowledge is legitimized is society. An awareness of these epistemological dynamics leads to new questions concerning the Western scientific faith in Cartesian-Newtonian foundationalism, as well as the certainty and ethnocentrism surrounding such a paradigm. In this context, those whose lives are predominately influenced by the "project of modernity" may come to understand that the knowledge-certification process is structured as much (or more) by socio-cultural and political factors than by a universal form of technical rationality.

The authors herein note that while it is necessary to create inclusive pedagogical strategies to assist all learners as they become literate in the content and processes of science, it is not sufficient. While challenging the boundaries of academic privilege and expanding the field of vision for science educators, the authors highlight the liberatory, political, and activist possibilities associated with creating a critical scientific literacy among all students. I believe that through this book you, the reader, will become inspired to respond to the call to action.

References

Apple, M. W. 1985. *Education and power*. New York: Routledge.

———. 1986. *Teachers and texts: A political economy of class and gender relations in education*. New York: Routledge.

———. 1990. *Ideology and curriculum* (2nd ed.). New York: Routledge.

———. 1992. Educational reform and educational crisis. *Journal of Research in Science Teaching, 29*, 779–789.

Carnoy, M., & Samoff, J. 1990. *Education and social transition in the Third World*. Princeton, NJ: Princeton University Press.

Dewey, J. 1931. *Philosophy and civilization*. New York: Minton, Balch.

Freire, P. 1983. *Pedagogy of the oppressed* (M. B. Ramos, Trans.). New York: Continuum. (Original work published in 1970)

———. 1985. *The politics of education: Culture, power and liberation* (D. Macedo, Trans.). Granby, MA: Bergin & Garvey.

Giroux, H. A. 1988a. *Schooling and the struggle for public life*. Minneapolis, MN: University of Minnesota Press.

———. 1988b. *Teachers as intellectuals: Toward a critical pedagogy of learning*. Granby, MA: Bergin & Garvey.

———. 1992. *Border crossings: Cultural workers and the politics of education*. New York: Routledge.

Giroux, H. A., & McLaren, P. (Eds.) 1989. *Critical pedagogy, the state, and cultural struggle*. Albany, NY: State University of New York Press.

Greene, M. 1995. *Releasing the imagination: essays on education, the arts, and social change*. San Francisco: Jossey-Bass.

Grumet, M. R. 1988. *Bitter milk: Women and teaching*. Amherst, MA: University of Massachusetts Press.

Habermas, J. 1972. *Knowledge and human interests* (J. J. Shapiro, Trans). London: Heinemann. (Original work published in 1968)

Harding, S. 1996. Science is 'good to think with.' In A. Ross (Ed.), *Science wars*. Durham: NC: Duke University Press.

Kyle, W. C., Jr. 1991. The reform agenda and science education: Hegemonic control and counterhegemony. *Science Education*, 75, 403–411.

———. 1999a. Science education in developing countries: Access, equity, and ethical responsibility. *Journal of the Southern Association for Research in Mathematics and Science Education*, 3, 1–13.

———. 1999b. *Critical issues of school and teacher education reform: Transforming science teaching and learning for a new millennium*. A keynote address at the Conference on Public Understanding of Science, Gaborone, Botswana.

Morrow, R. A., Torres, C. A. 1995. *Social theory and education: A critique of theories of social and cultural reproduction*. Albany, NY: SUNY Press.

Pinar, W. F. (Ed.). 1988. *Contemporary curriculum discourses*. Scottsdale, AZ: Gorsuch, Scarisbrick.

Pinar, W. F., & Reynolds, W. M. 1992. *Understanding curriculum as phenomenological and deconstructed text*. New York: Teachers College Press.

Stanley, W. B. 1992. *Curriculum for utopia: Social reconstruction and critical pedagogy in the postmodern era*. Albany, NY: State University of New York Press.

Touraine, A. 1995. *Critique of modernity*. Cambridge, MA: Blackwell.

Ødegaard, M., & Kyle, W. C., Jr. 2000, April. *Imagination and critical reflection: Cultivating a vision of scientific literacy*. A presentation at the annual meeting of the National Association for Research in Science Teaching, New Orleans, LA.

Introduction

Angela Calabrese Barton
Margery D. Osborne

This book is devoted to examining marginalized discourses (critical, feminist, multicultural, and poststructural theories) and the implications these have in considering the ideal, a science education for all children. These perspectives are brought together because they combine a questioning of the foundational canons composing science as a discipline and science education as a practice with an understanding that the intersections of race, class, and gender, and other forms of identity labeling, frame access to knowledge and power. Our desire to develop such a book stems from our own experiences in which we have been involved in researching science teaching and classrooms, endeavoring to address the multiple needs of girls and minority students. We have found this work to be both complex and infused with problems and dilemmas which do not lend themselves to simplistic, prescriptive solutions. Indeed, the papers presented in this book all highlight the complexities of teaching science in a diverse culture and how these complexities suggest rethinking foundational assumptions about pedagogy and the discipline of science.

In April 1998, we edited a theme issue of the *Journal of Research in Science Teaching* focused on marginalized discourses and science education. The purpose of that particular theme issue was to introduce to the science education community questions and concerns raised by the lives of marginalized groups such as girls, children in poverty, ethnic and racial minority children, and language minority children.

The studies presented in the theme issue drew from still current debates concerning schooling and the need for liberatory education, the social construction of science and of identity, and systems of race, class, and gender oppression and domination. These works force us to confront such questions as

- How can historically marginalized students become involved in science?
- How can we shape practice and curriculum to address the needs of diverse learners?
- How does reshaping practice and curriculum alter our thinking about the discipline of science itself?

Since that time, we have been asked to talk more about these questions and about how theories such as critical theory or feminist theory really help us to understand and respond to all children better and what it really means, in the day-to-day practice of teachers, to enact more liberating pedagogies. We begin to answer some of the questions in this book. In fact, this edited text builds on our work begun in the theme issue of the *Journal of Research in Science Teaching* in bold ways. We have asked difficult questions of what happens when we open our science classes and our science teacher education classes to students' lives and experiences as sources of science and of critique. In other words the book is about what can happen when these theories are applied in science classrooms. For example, in what ways do the marginalized discourses in theory and practice push us to fundamentally reformulate our conceptions of teachers/teaching, students/learning, subject matter knowledge (science and what it means to know and do science), and the multiple relationships between and among these domains? In what ways do the marginalized discourses in theory and practice push us to fundamentally reformulate our conceptions of science for all?

This book serves two purposes: to educate readers about the importance, history, and possibilities for marginalized discourses in science education, and to engage readers in multiple cases where contributors have systematically applied and examined what happens when these theoretical frames are brought to bear in classroom practice (prekindergarten through twelfth grade science and science teacher education). Each chapter will draw from both theory and classroom practice (where "classroom" is broadly construed to include any teaching and learning setting). Many of the authors will speak from the voice of teacher-researcher; however, this will not be the case in all of the chapters. For example, all of the authors have attempted through their teaching and research to deconstruct the canon of science as well as critique who one must be to partake in that canon. All of the authors have attempted to situate and problematize knowledge construction about science and self within the everyday—a framework that is marked by discourses of domination, control, opposition, resistance, and power. They show us the ways in which these discourses emerge from pedagogical encounters and frame teachers, students, and classrooms. By arguing that teachers and students need to combine a critical understanding of science, including its content, culture, and discursive practices, with an understanding of students and educational processes, we can develop a deeper appreciation of how students and teachers must use these understandings as a basis from which to enrich the ideal of a science for all.

Structure of the Book

The first chapter introduces the reader to the primary questions raised by marginalized discourses in science education described above. Critically examining the assumptions undergirding the reform initiatives in science education, we argue that central to rethinking science education for all is the idea of responsive teaching in a political context. In other words, science education ought to be about more than passing on the disciplinary knowledge of science. Science with children is an incredibly complex social site marked by multiple interacting layers of power arrangements and social and institutional forces which shape and define the boundaries of what is possible. This suggests just how much science education is about issues of power and relationships on a personal as well as an institutional level. As K. Weiler claims, "teaching extends beyond subject matter knowledge; the centrality of teaching lies in a recognition of the values of students' own voices, subjective experiences of power and oppression, and the worth of their class and ethnic cultures" (*Women Teaching for Change* [South Hadley, MA: Bergen and Garvey, 1988], 148). Envisioning science as a social construction locates the learning and doing of science within social relationships on a day-to-day level, not just historically. Allowing these relationships, and the questions and concerns they provoke, to guide the science creates opportunities for self- and social empowerment.

The remainder of the book is divided into two parts. Part 1 (chapters 2 through 6) focuses on elementary education and elementary teacher education. In chapter 2, Osborne and Brady examine elementary science through the lens of magical realism. Using this stance, they argue that fundamental to transforming science education for children is learning to see children, their work, and ideas through new patterns and new metaphors and learning; to allow the mystical and the magical to open up ways of understanding and modes of expression in responding *with* children in classroom contexts. In chapter 3, Bulmahn Barker builds on the argument presented by Osborne and Brady regarding the power of building new spaces for teacher thinking. Using Virginia Woolf's idea of a "room of one's own," Bulmahn Barker presents a case for teachers needing to have the freedom and space to think about teaching practices and make these their own. This approach to curriculum and pedagogy changes emphasizes the understanding of reform efforts and is based on the belief that there is no single effective method for teaching and learning as students and teachers differ in their preferences, desires, imaginations, attitudes, and interests.

In chapter 4, St. Louis and Barton use the stance of teacher-researcher in an elementary science school located on site at a homeless shelter to present a series of dilemmas regarding how difficult it can be for teachers to challenge and remake their ideas regarding science, children, and teaching. Using these dilemmas, these authors make the case that if science education is going to be liberatory for teachers and students alike, then science teaching must be emergent, and teachers must work hard to promote reflexive transformations among themselves, science, and students. In chapter 5, Zembylas adds to the embedded but heretofore implicit dimension of the developing conversation in this book. He presents an essay about the value of elementary school teachers' emotions in their teaching of science. Finally, in chapter 6, using a feminist lens, Howes examines what it means to work with preservice elementary teachers in engaging the kinds of critical consciousness-raising presented in chapters 2 through 5 in order to construct a teaching practice really aimed at science for all. She argues that we must begin to use the strengths of preservice teachers rather than their weakness (i.e., not enough content knowledge, afraid of science, etc.). She shows how elementary preservice teachers' enthusiasm for hands-on science instruction and their conviction that science is "everywhere" can serve as that crucial starting point.

Part 2 (chapters 7 through 12) focus on secondary education. In chapter 7, Hildebrand shows how the discourses that re-present science through writing in secondary schooling are built upon foundations that can be legitimately challenged from several critical standpoints. Deliberately breaking the conventions of academic discourse in the framing of this chapter, through such tools such as poetry, she presents the ways that some teachers are rewriting science with students in schools. In chapter 8, Fusco also takes up the question of representation in secondary science. Drawing from her teacher research with homeless youth in after-school programs as well as critical/feminist theory, she deconstructs the power-control-knowledge nexus in assessment in science education and argues for representing science through performance *as* assessment. In chapter 9, Weinstein focuses the question of representation of science in secondary education through "guinea pig science." This chapter demonstrates the importance of multiple, positioned narratives in science education. Weinstein argues that science teachers and science education researchers should take into account, specifically, the perspective of human research subjects (and research subjects more generally) within their pedagogical and scholarly works. In making this case, Weinstein carefully weaves together a brief overview of guinea pig history and politics with a discussion of feminist standpoint theory, an approach to knowledge which helps theorize the possible role and importance of guinea pig (and other marginalized) points of view in the science curriculum.

Bianchini and Cavazos move us into a discussion on secondary teacher education. In their chapter on professional development (chapter 10), Bianchini and Cavazos critically examine from a feminist, multicultural stance their efforts to support practicing science teachers and university scientists interested in integrating inclusive content and pedagogy into their educational practices. Drawing from individual and group interviews, they present four case studies of teachers and scientists who have worked to implement both inclusive science content and instructional practices in their courses. From these cases, they call for professional developers to provide teachers with multiple strategies for locating and integrating inclusive science content; opportunities to work collaboratively with colleagues; on-going, sustained professional support; and, most importantly, the freedom to adapt ideas to local circumstances. Such steps, they argue, are necessary if professional developers are to aid teachers in making inclusive science education a reality in schools.

In chapter 11, Parsons uses feminist autobiographical inquiry to challenge the truth regimes which dominate science and university education. In uncovering these regimes she points us towards two challenges: the need to deal with the sacred stories of university practice, stories that contribute to maintaining the established power-knowledge relationships from multiple feminist perspectives; and the need to deal with the sacred stories of theorized feminism, stories that have informed her on-going analysis. She uses these dilemmas to argue for ways in which university science educators might better work with teachers to uncover the regimes of truth guiding schooling and science at all levels of teaching. Finally, in chapter 12, Rodriguez presents us with a call for action. In his essay, he proposes that we urgently need to (re)examine our roles as researchers and teacher educators in order to effectively address the reproduction of social and educational inequalities in our schools. Using the concept of cultural warriors, Rodriguez argues that what we need to do is muster the courage to manage the resistance and the risks associated with helping implement our proposed changes, have the courage to escape the entrapment of our own privilege in order to take a step closer to understanding how our actions and inactions make us either pawns for the status quo or effective agents for transformative social justice, and expand our gaze as researchers and teacher educators and rise to the challenge of becoming cultural warriors for social change.

Chapter 1

Marginalized Discourses and Pedagogies: Constructively Confronting Science for All in Classroom Practice

Angela Calabrese Barton
Margery D. Osborne

My name is Cynthia and I am nine years old. I am in the fourth grade. I have lived with my family in Austin, Texas, for five years. Before moving to Austin I lived in Mexico. Most of my family still lives in Mexico but some of my family lives here in Texas with me. Right now my family lives in a homeless shelter. We have lived here for one year. Although I have lots of friends to play with here, I don't like living here when it gets dark outside. It is not safe. I also do not like it when boys think they are better than me, or when they think I cannot do all of the same things as them! I only like school sometimes. I like school when we get to party and say whatever we want. I also like school because I can see my friends. I do not like school when my teacher is mean to me, which is most of the time. I don't like science and I definitely hate art! My teacher always is picking on me.

My name is Jessica and I am eight years old. Cynthia is my best friend. I also go to school in Austin, Texas. My parents are originally from Mexico, but from a different part of Mexico than Cynthia's family. I live in the same homeless shelter as Cynthia. Our families moved into the shelter at about the same time. I really like spending time over at my friends' places, but I don't it like when people fight especially when people fight in my face! I also don't like it when boys try to show me up. I can run as fast as any of the boys around here! I also don't like going outside at night because it is not safe around here! I never like school, mostly because it's boring. It's boring because we have teachers and work, and we never play. I always have to be here, be there, do this, do that, and I am always getting in trouble. It's not even my fault, and I get in trouble. If schools did not have teachers, then it might be fun.

We want to tell about learning science in school. Actually, when Angie asked us to write a story with her, we did not want to do it at first because it was supposed to be about science in our school. The truth is, we really do not like science. Our favorite subjects (if we had to pick one) are math and reading.

Well, we have come up with a story, and we are going to call it "el secreto de las ninas" because the reason we do not like science is that we are not used to science! Sometimes when we complain about science in school, people think we don't like it because we are girls and that is not true. They tell us "we can be anything we want to be!" We know that, sort of. We are just as good as boys, we just don't want to explain our reasons. It is our secret.

We find the girls' secret interesting, but what do they mean? What does it mean to be used to something? Does it mean that their schoolteacher never teaches science at all and so they are not used to having science class at all? Does it mean that when their teacher does teach science the girls are not used to the topics or the content or the pedagogical methods the teacher uses to teach science? Does it mean that they are not used to the language or the skills the teacher requires them to use? Or, does it mean something entirely different from the ideas we suggested? We asked the girls.

Okay. Let us tell you a story about what happened in school to show you what we mean. Our teacher told us we were going to start a unit on "movie making." We both thought this would be really cool because we love to go see the movies! One of the first activities we were going to do was make a shoebox camera. We would make this camera out of a shoebox and it would take real pictures. And then we would be able to keep the camera and take whatever pictures we wanted! At the beginning we were really excited about this project because neither of us have ever owned a camera before. We remember only one time when we were able to use a disposable camera because Cynthia's mother bought her one from the store for her birthday.

The day before we were to start the camera project our teacher asked each student to bring in an empty shoe box. Well, where are we going to get empty shoe boxes? I told my teacher I didn't have an empty shoe box so then she said to the whole class, "If you don't have a shoe box, you can ask your mother or an older brother or sister to take you to a discount shoe store and ask for a shoe box. They will give you one for free." Then she said, "For those of you who cannot get your own shoe box, you can bring in fifty cents and buy one from me." She then told us that we "are getting older now and we have to learn to become more responsible for our own education." Well, we are still like, where are we going to get a shoe box? My mother cannot take me to the shoe store! She doesn't even have a car, and she cannot speak English that well, and she has to watch my baby brother. And, then we are also like, where are we going to get the fifty cents? So, the next day we went to school without our shoe boxes and without fifty cents.

At the beginning of the day the teacher collected the shoe boxes. A couple of other students in the class did not have shoe boxes or money either. Our teacher gave our class a lecture about being responsible. So, we told her in private right before recess that our mothers could not take us to the store and we did not have any money and that is why we did not have a shoe box. She asked us why we didn't tell her earlier. She also told us it was okay, and that she understood. She told us we could help clean the erasers during recess to "earn" the shoe box and that it would be "our secret." Well, we both decided to go to recess because we were mad at her and we didn't want to share a secret with her. When it came time for science our teacher said nothing to us, but gave us a shoe box anyway so we could make our camera. But we were the last ones to get a shoe box, and they were ugly. By that time we did not want to make a camera anymore, and we just sat there and poked at our boxes.

We begin with this story narrated by Cynthia and Jessica because we believe their story raises questions about how science, power, and privilege intermingle in the context of learning and doing science. And their story does so on many levels. On an immediate level, we wonder:

- What do the girls really mean when they say that they are not used to science?
- Whose experiences are valued in this setting and in what ways?
- What does it mean to engage all learners in science?
- How should a teacher handle a situation like the one described by Cynthia and Jessica?

On a more generalized level, we wonder:

- What does it mean to teach science in a way that values the lived experiences, ways of knowing the world and social identities held by all students, especially women and minorities?
- How do students' concepts of science constrain roles and expectations, shaping power and privilege in science class by defining roles and identities?
- How can we move beyond these assumed roles and perceptions to help all children become empowered and liberated through developing science understandings?

These questions are central to contemporary science education efforts. After all, recent reform efforts in science education suggest that all students should attain some foundational knowledge of the substance and processes of science. Encapsulated by the phrase science for all, these are described in the report "Project 2061" (AAAS 1993) and subsequent documents (Goals 2000 1994; NRC 1996). Such an ideal, however, fails to recognize or address the implications in defining such a canon or enabling its acquisition by students under either a version of society in which canonical knowledge is integral to its functioning or one in which knowledge is the foundation upon which change is enacted. In other words they do not ask the primary curriculum questions, Who's knowledge? or Knowledge for what purpose? Visions of science for all and derivative articulations of science education are inherently conflicted because they don't ask hard questions about the sources or functioning of such knowledge and they sit on both sides of ideas concerning the function of knowledge in society without acknowledging them.

Thus, in this first chapter we do three things. First, we examine the ideal science for all as it has been put forth by contemporary reform documents and reform-minded research and practice. We use this examine to highlight the major challenges faced by those committed to understanding and enacting empowering and equitable education for all children. Second, we look towards marginalized discourses in science education to better understand the challenges we lay out. To do so, we first lay out, in a comprehensive fashion, what we mean by marginalized discourses and what these discourses have to say to science education. We then apply these understandings to the critiques levied on reform-based science education. Third, and finally, we develop an argument for political-responsive science teaching as an image of what we mean when we say marginalized discourses and science education. We hope this image clears a path for making sense of the chapters which follow.

Science for All?

According to Eisenhart, Finkel, and Marion (1996), the US reform initiatives view scientific literacy for all Americans as the educational solution of four problems thought to restrict or even prevent an adequate preparation in science for the next century: (a) there is a low level of scientific knowledge among members of the population, (b) few citizens are prepared to use scientific knowledge to make decisions that affect their lives, (c) the percentages of women and minorities in many science fields remain disproportionately low, and (d) science is said to be poorly taught in schools. Thus, both the American Association for the Advancement of Science (AAAS) and the National Research Council (NRC) in the US urge the nation to make scientific literacy for all the overarching goal of science education reform: "The goal of the National Science Education Standards is to create a vision for the scientifically literate person and standards for science education that, when established, would allow the vision to become a reality" (NRC 1996, 1). "When demographic realities, national needs, and democratic values are taken into account, it becomes clear that the nation can no longer ignore the science education of any students" (AAAS 1989, 214).

As might be expected many recent reform initiatives globally have supported scientific literacy for all: "All students, regardless of gender, cultural or ethnic background, physical or learning disabilities, aspiration, or interest and motivation in science, should have the opportunity in science, should have the opportunity to attain higher levels of scientific literacy than they currently do. This is a principle of equity...and has implications for program design and the education system...to ensure that the standards do not exacerbate the differences...that currently exist" (NRC 1996, 6–7). "When demographic realities, national needs, and democratic values are taken into account, it becomes clear that the nation can no longer ignore the science education of any students. Race, language, sex, or economic circumstances must no longer be permitted to be factors in determining who does and who does not receive a good education in science, mathematics and technology" (AAAS 1990, 214).

The science for all reform effort has been hailed by contemporary science education researchers as critical to the education of women and minorities (Anderson 1991; Rosebery 1994). The reform stresses the importance of making the rules, structures, content, and discursive practices of science explicit and accessible to all students with the direct goal of creating a scientifically literate citizenry. In this framework, to be scientifically literate means that one is able to "grasp the interrelationships between science, mathematics and technology, to make sense of how the natural and designed worlds work, to think critically and independently, to recognize and weigh alternative explanations of events and design trade-offs, and to deal sensibly with problems that involve evidence, numeric patterns, logical arguments, and uncertainties" (AAAS 1993, xi). In one sense this reform effort mirrors the egalitarian tradition in American democracy by claiming that all citizens have the right as well as the responsibility to be aware of scientific concepts, and this reform should enable a progressive vision of democracy, a vision in which all voices count. It differs from traditional versions of science teaching, versions based more on finding and educating future scientists.

However, we argue that although such a vision of science for all might be important, it operates on three assumptions which should be addressed if a science for all is to be truly constructed.

- First, "Project 2061" (AAAS 1993) draws strength from a belief that schools are meritocratic in nature. Schools have played a historical and social role in reproducing race/ethnicity, class, and gender inequalities in part through disciplinary studies and pedagogy (Anyon 1984; Harding 1991; Reyes and Valencia 1993). Documents like "Project 2061" perpetuate the illusion that if all children learn science, all children will be equal.
- Second, the reform maintains a deficit model of minority knowledge: minorities are *lacking* in important knowledge. By favoring only white, male, middle-class cultural values, it implies that minorities and women are inferior (Apple 1992).
- Third, it assumes that students will choose to adopt these values when their own are shown inconsistent and (implicitly) inferior. When students do not, it is assumed that they are at fault *not* the instruction or the content of instruction (Foley 1991; Apple 1979).

These three assumptions neglect to acknowledge that science and its practices reflect power differentials in our society and that science education is nestled in the politics of assimilation and meritocracy (Barton and Osborne 1995). The politics of the construct of difference is a legitimization of dominant society under such assumptions. Acknowledging these assumptions allows a potential for revisioning both science and school practices. Without acknowledging these assumptions the reforms require minority students to silence their cultural and linguistic heritage and to embrace a way of knowing which has effectively defined minorities and women as socially and intellectually inferior (Davidson 1994). The US science education reform initiative, "Project 2061," states that "teachers should . . . make it clear to female and minority students that they are expected to study the same subjects at the same level as everyone else and to perform as well" (AAAS 1989, 151). This message implies that minority students and females need to work and act like their white male counterparts not that either science or instruction will be modified to accommodate them. Although it can be argued that this is a call for teachers to engage all students, not just the white middle-class males, in the academic rigors of science, it can also be read as a call for teachers to encourage, if not require, acculturation. In the very effort to create inclusive science education communities, policy, practice, and curriculum become connected in the politics of assimilation with schools and teachers as agents (Calabrese and Barton 1995).

It seems to both of us that we need to ask an extremely important question: Who are we thinking about when we dream of a science for all? What is a science for all like? Wouldn't a science for all look different from the education we are now trying to enact? We are suspicious that many people see a science for all as involving an all that becomes increasing homogeneous. "All" is not a word that carries heterogeneity. It suggests, instead, likeness and similarity. The children who are "different" slowly becoming more like all of us (whoever we are). We would like to pose an argument that answers to such questions do not involve thinking of ways to enable marginalized students—or any students—to engage in present educational forms. Rather, an education for marginalized children involves rethinking foundational assumptions about the nature of the disciplines, the purposes of education, and our roles as teachers. It does not mean remaking those children into our own images. It involves remaking schooling and science in their often multiple images.

In our society the question of what to do with difference in our classrooms has been a perennial one. As teachers we know that every child is different, behaviorally and in background, interests, and ability. Sometimes we celebrate that difference but other times that difference is an impediment—it gets in the way of our teaching (Ball 1993; Lensmire 1994; Osborne 1997). A central question in teaching, extending from our worrying generically about how to get through another day to our pondering how are we going to teach particular science concepts, is, What do we do about difference? Such a question should be similarly important when envisioning the construction of a science for all.

Critical Science Education

Reform initiatives in science education in the 1980s and 1990s have squarely positioned science as a social process and cultural practice with particular ways of knowing and doing science (AAAS 1993; NRC 1996). In the last decade there have been numerous research articles which draw from theoretical traditions (critical, feminist, multicultural, and poststructural theories), as well as from these reform initiatives, to challenge the positivistic foundation of science and school science as a basis for understanding issues of access, equality, and excellence in science and science education. Although each theoretical tradition uses a different analytic lens (i.e., critical theory: class; feminist theory: gender), all of these traditions raise fundamental questions of power, knowledge, and production in science and schools. The driving goal of these efforts has been to construct images of a more inclusive science education, whether exclusivity is defined in terms of race, class, gender, or other marginalizing labels or identities.

Drawing, in a comprehensive fashion, from these "critical" traditions in science education, we have attempted to cull together the common threads. Namely, critical science educators have pushed the debate surrounding inclusive science education forward in terms of how we understand the nature of science and knowing in science, the relationship between science, schooling, and society, and the implications these belief structures have for how we view school science. The questions emerging from this debate are, for example, What implications does a critical perspective have for teaching an inclusive science across issues of diversity? What lessons can we draw from such a perspective as we seek to strengthen the impact of performance assessment on equity in science education? In what follows we begin to answer these questions by examining these three domains from the perspectives of critical science education. We then use the insights gained from these perspectives to specifically address equity and diversity in assessment and describe our own attempts to imagine a performance assessment that is both a means for understanding the enactment of critical science education as well as a method for rethinking the nature of performance assessment in science.

The Nature of Science and Knowing Science

Critical science education draws from the feminist and multicultural belief that science is a subjective but rigorous and reflexive approach to making sense of (and building stories about) the world. Critical science education views scientific knowledge as constructed through social acts where individuals interact in distinctive ways with society and culture to create something for some purpose (Gill and Levidow 1989). In other words, the production of scientific knowledge is linked to the social uses of and needs for scientific knowledge (Harding 1998; Young 1989). Critical science education, therefore, reasons that the knowing and doing of science are historically, socially, and politically situated processes. What scientists know and how they have come to know it are artifacts of the context in which scientists work.

For example, in evolutionary studies, a great deal of attention has been placed on interactional behavior in its relation to the development of human anatomy (Harding 1986; Hubbard 1990; Longino 1990). It has been studied in great detail as a potential argument for biologically determined sex roles (Harding 1986). The studies, according to Harding "show a high tendency to project onto ape nature and social relations both racist and sexist projects of the observer's own society" (1986, 96). They have been used to justify and perpetuate masculine dominance and restriction of women's opportunities (Harding 1986, 83). Thus, Harding asserts, "androcentric assumptions, then, appear in the collection, interpretation, and use of the data" (1986, 96). She argues that even though this is a blatant example of value-filled science, there are many more subtle ones that probably never get picked up on because of the deeply embedded value system in our daily lives.

In order to eliminate androcentric science, Harding (1986, 1991) contends that the discovery of gender, its individual, structural, and symbolic consequences that account for woman's oppression, provides a lens through which scientists can and must view the world. A woman's experience is as equally valid a resource as a man's: Scientific inquiry must also begin with questions that originate in women's experiences. Furthermore, women's perspectives on their own experiences provide important empirical and theoretical resources for research. "They generate research problems and the hypothesis and problems that guide research. They also serve as a resource for the designing of research projects, the collection and interpretation of data, and the construction of evidence" (Harding 1989, 28). Until gender is recognized, science will remain gender exclusive.

From another perspective, Steven Lubar (1987) has described how a range of cultural values and technical ideas can be embedded in a single machine. Using the case of John Howe, who designed an automated pin-making machine in the 1830s, Lubar revealed how common mechanical movements, the social structure of the pin society, the skills of the machines' builders, and prevailing design practices gave Howe's machine a distinct style.

Haraway's (1997) studies of genetics and cyber technologies also demonstrate how science is an ongoing interaction among "core narratives" shaped by political, economic, and cultural contexts that reflexively guide observations, theory building, and applications. For example, Haraway describes her (fictional) historical, moral, physical, and intellectual relationship with the first genetically engineered, patented organism (Oncomouse) as well as with other equally provocative (yet often under-politicized) domains of society such as scientist-activists, schoolteachers, and vampires. In describing these relationships she concludes that all knowledge is situated, partial, and context dependent and that the only way to move towards a more holistic and difference-making knowledge base is to generate multiple understandings from different contexts. These differing contexts must include those from within and without the scientific community, including such individuals and groups as medical patients, the homeless, teachers, artists, and construction workers, to name a few. The intersections of these understandings (and contexts) will help to provide new and different ways for making sense about and acting in the world.

As the Haraway example suggests, scientific agendas are informed by a community greater than just scientists, and such a vision neglects the day-to-day practice, struggles, and meaning-making of scientists and the situationally contingent understandings and agreements that scientists construct. For example, as a direct result of AIDS community activists learning the science and scientific discourse around HIV/AIDS and then communicating their own needs to those bodies in the language of science, government agencies such as the National Institute of Health have revised their guidelines for clinical trials, double blind trials, and control groups. Additionally, doctors and researchers have revised their research plans not only to take into account the needs expressed by those with AIDS, but also to refine their methodologies to deal more compassionately and humanely with AIDS sufferers. Thus, scientific research is influenced by the overall research context, the specific research situation, and the situated ways in which scientists act, think, and work converge with the descriptions of ordinary people (Lave 1988; Roth and McGinn 1999).

One outgrowth of this perspective regarding the nature of science and knowing science is that scientific knowledge is viewed as tentative and imbued with the values of the individual and the culture in which it was generated. This does not make scientific knowledge any less useful according to a critical science perspective. Rather it simply helps to state more explicitly its limitations. Another outgrowth of this perspective is that one can never know or do science separate from his or her own history (individual and societal). Although we can try to understand our own history and how it might influence how we come to know the world, these two can never be fully separated. Such a perspective about knowing and doing in science is in contrast to the historically accepted vision of science as an objective enterprise.

Intersections between Science and Society

In addition to understanding and questioning the nature of science and scientific knowledge, a critical science education perspective suggests that it is important to make visible how science is situated within larger social values and global ecosystems. Scientists' aim, since the seventeenth century, has been the control and the domination of nature (Keller 1985). This authoritative and controlling stance has helped catapult science into the category of "invincible." Keller (1985) equates this power picture with paranoia and goes on to suggest that the need to have "the only interpretation" is equivalent to the need to measure one's own strength against another's submission, or a dream of the dominion of science over nature. Feminist and multicultural scholars, such as Keller, suggest that science should not be an exercise of domination, but rather one of equity (Gill and Levidow 1990).

In addition to taking part in the societal power pyramid, traditional science has its own internal power pyramid based on competition, capital, and control. Longino (1989) suggests that science labs are typically structured hierarchically and that scientists relate to one another through competition. These two features of traditional science—the hierarchical organization of scientific knowledge and scientists within society and within the discipline, and competition—are not necessarily a feature of science, but rather features of Western society to which Western society has grown accustomed (Harding 1998).

Finally, science and science education are situated within their representations of the natural world and their set ways for regulating meaning. This situatedness is central to understanding how dynamics of power and privilege structure the daily life of society. Scientific concepts emerge from dealing with societal problems/real life and the needs of the local community, needs which are seen as fundamental to the creation and production of science. Viewed from another angle, this can also be read as science as a social practice with social responsibilities (Epstein 1997).

That is, critical science underscores the stance that science and scientists do not have control over nature, rather they have an ethical responsibility for the knowledge it produces about the world. For example, Cynthia Cohen (1996) describes how the development of reproductive technologies in the United States, and in particular the process for egg donation, is inseparable from the needs of the mother, the donor, and the doctor, as well as the ethical and moral implications in a technologically advanced and socially conservative society.

A detailed look at writings around scientific invention helps to make this case about science and society. Traditional biographies of inventors have drawn primarily on a modernist rendering of invention, where "to invent" is defined as "to discover that we know not, and not to recover or resummon that which we already know" (Bacon 1963, 268–269). A modernist discourse of invention makes certain claims about ownership. Inventions are often described through patents and trademarks. When an individual applies for a patent, or to claim intellectual rights or ownership for an idea discovered, he or she must textually document that their invention satisfies the following three criteria: Is it new; i.e., has anyone else already discovered or created this or something extremely similar? Is it useful? Is it unobvious? (Casey 1997). Patented inventions are then required to be named or labeled, thereby signifying that there is some fixed meaning to the invention and that the inventor has some sort of control over it (Keller 1985). In fact, trademarks developed so that individual inventors did not have to reveal the "secret knowledge" behind their invention. The recipe for Coca Cola is an example of one of the most famous trademarked "secret inventions."

However, Gorman and Carlson (1990) and Bijker (1987) in science and technology studies, Foucault (1975) in postmodern literary critique, and Haraway (1997) and Krikup and Keller (1992) in feminist theory have all critiqued the traditional rendering of inventions. They argue that the modernist analysis of invention makes it hard to understand the inventive act because these technological and biographical accounts have relegated invention "to the realm of mysterious genius" where invention can only be described but never explained (Gorman and Carlson 1990, 132). Relegating invention to the realm of mysterious genius masterfully conceals invention as social act. Furthermore, describing invention solely within the realm of technology, rather than within the continuous and shifting terrain of science/technology, shrouds the connections between the unstable, socially contextualized, and recursive process of invention and invention as embodied agency.

Science/technology and inventive acts are inscribed by individuals interacting dialectically with socioculture in a distinctive way to generate something (Foucault 1975; Haraway 1997; LeFevre 1987). Even when the agent is a single individual, invention is still a social act because the self is socially influenced (Weber 1949). Human agents always act dialectically—in the contexts of their interconnections with others and with the socioculture (Buber 1970; Geertz 1973).

Invention as a social act raises questions about the nature of ownership. For example, in writing about literary invention, Foucault (1975) argues that the concept of author as one who authorizes or invents a text is of relatively recent origin; that there was a time when poems, narratives, and plays were composed and circulated without the question of authorship ever being raised (D'Angelo 1987). This raises the questions, What is an author? And, can text be invented? According to Foucault, although people author or invent the narrative or the poem, they do so through the sociocultural framework: "One comes to the conclusion that the author's name does not refer to a real person but that it exceeds the limit of the texts, that it organizes them, that it reveals their mode of being, or at least characterizes them. Though it clearly points to the existence of certain texts, it also refers to their status within a society and within a culture. . . The function of an author is thus characteristic of the mode of existence, circulation, and operation of certain discourses within society" (cited in D'Angelo 1987, x). Foucault is suggesting that the question of ownership in invention ought to be considered unbounded and unstable because ideas are developed and enacted through a broad sociocultural network. Feminist scientist Keller supports this point; she argues that to understand science, one must focus on "the personal, emotional, and sexual dimensions of the construction and acceptance of claims" (1985, 9). Similarly, Longino (1990) argues for science as social knowledge because it embodies context-laden process.

Science as a School Subject
Currently, the means being used by the AAAS and the NRC to promote scientific literacy for all Americans "[specify] what facts, concepts, and forms of inquiry should be learned and how they should be taught and evaluated" (Eisenhart, Finkel, and Marion 1996, 266). In fact, there is widespread agreement, among feminist, critical, and multicultural science educators, that students need to have access to the domain of Western science, even if it has little relevance to the lives of students most on the margins of school science (Atwater 1996). However, such access must

occur in ways that are culturally relevant to students. For example, in traditional science classrooms, contradictions often exist between the unproblematic way in which science is presented and the ways in which students' gendered, raced, and classed values are a part of their own construction of science (Atwater 1996; Barton and Osborne 1998; Brickhouse 1994; Lee and Fradd 1998; Rodriquez 1998). These contradictions, because they are often unarticulated and unrecognized, teach students that if standard ways for engaging in science do not make sense, feel right, or connect to their experiences, then they are the ones who are wrong or intellectually deficient. Students are expected to make sense of the world in prescribed ways; they learn to impose boundaries, constraints, and definitions on themselves, others, and the world. In short, they learn that a lack of diversity in the ideas and ways of knowing is what is acceptable in science.

Critical school science researchers also argue that the culture of minority students often conflicts with the philosophy supporting US reform proposals or the culture of schools (Brickhouse 1994). According to Atwater (1996), the relationship between science and the children from diverse cultures and languages is problematic because school science typically reflects middle-class experiences and excludes the lives of students most on the margins of school science. This last point has been developed in rich detail by the work of Lee and Fradd (1998), who have demonstrated the importance for teachers to build instructional congruence between the languages and cultural experiences of second-language learners and those central to success in science. In short, maintaining what scientists do without considering whom one must be to do science and what school science often includes/excludes does not create spaces where multiple perspectives in knowing and showing in science can emerge.

Thus, it is not enough to teach students rules for participation in science if those rules do not connect to, and perhaps even conflict with, the students' out-of-school lives (Ladson-Billings 1994; Rosebery, Warren and Conant 1992; Rosser 1990). Critical science education supports pedagogical strategies that build with the ways of knowing brought to school by students, ways such as caring, cooperation, and holistic approaches, even when those ways of knowing are not obviously part of science (Barton and Osborne 1999; Roychoudhury, Tippins and Nichols 1995). Also supported are strategies who seek to incorporate communication processes reflective of the lives and cultures of students which may be present in the classroom, processes such as oral narratives and storytelling (Atwater 1996).

Furthermore, Atwater (1996) argues that science educators need to use—and help their students to use—a critical lens to question how scientific knowledge is learned and produced, and the ways in which classroom practice links (or does *not* link) science education to self and social empowerment. Most critical science educators agree that science and its uses in society ought to be critiqued and challenged, although the level of critique is contested within feminist, critical, and multicultural science education circles. Some argue that it is the teaching methods and the applications of science that ought to be challenged—not necessarily the underlying scientific concepts and principles (Lee and Fradd 1998), whereas others argue that the underlying science needs to be challenged as well (Gill and Levidow 1989; Helms 1998). From this latter critical science perspective, teaching and learning positions students in an articulated relationship where both science and students can change; it is a two-way relationship. Students should be users and producers of science and should develop the understandings and habits of mind that they need to become compassionate and informed human beings (Eisenhart, Finkel and Marion

Table 1.1 Comparative Summary of Critical Science Education

	Traditional School Science	**Progressive School Science**	**Critical School Science**
Nature of Science and Ways of Knowing Science	• Positivist world view—scientific knowledge is an objective representation of how the world works. • Concepts are really explanations of natural phenomena to be mastered for tests.	• Scientific knowledge is what scientists produce. It is an objective representation of how the world works. • Concepts are really explanations of natural phenomena. It is interested in conceptual understanding.	• Scientific knowledge is a human-made explanation of how the world works and is thus subjective yet rigorous and reflexive. • Concepts, although rigorously tried, are culturally based and need-based explanations of natural phenomena to be applied in everyday life activities. • Science is a social activity and involves understanding how human values and characteristics shape scientific knowledge and understanding.

Table 1.1 Continued

	Traditional School Science	Progressive School Science	Critical School Science
School, Science, and Society	• Scientific concepts are seen as outcomes in and of themselves. • Positions students in a relationship where only they can change—one way relationship.	• Scientific concepts are seen as a commodity for dealing with in-class science problems and scripted human applications. • Positions students in a relationship where only they can change—one-way relationship.	• Science has an ethical responsibility for the knowledge it produces about the world. • Scientific concepts emerge from dealing with societal problems/real life and the needs of the local community, needs which are seen as fundamental to the creation and production of science.
Goals and Purpose of Science Education	• Organized and tested knowledge must be "transmitted" through lecture and strict experimental procedures. • Students have to memorize scientific knowledge and procedures within the structure that was established. Science is for the scientists.	• Organized and tested knowledge interconnected with mathematics and technology. • Students must develop the understandings and habits of mind they need to become compassionate and informed human beings.	• The teaching and learning of science ought to contain elements of action and change, i.e., learning is not just an academic task; it is about interacting with/in the world. • Science teaching and learning should include the content, process, histories, norms for participation, and discursive practices. • Students should be viewed as users and producers of science.

1996; Rodriguez 1998). By engaging students in activities that connect science to society and community in authentic, useful and needed ways, scientific literacy can be more relevant and less exclusive than it is under the current reform agenda (Eisenhart, Finkel and Marion 1996).

Finally, critical science education rejects the idea that science teachers are simply transmitters of existing configurations of scientific knowledge.

Teaching science cannot be reduced to the acquisition or mastery of skills or techniques but must be defined within a discourse of human agency. The teaching of science occurs within the larger contexts of culture, community, power, and knowledge. Science teaching therefore must respond to the political and ethical consequences that science has in the world, and must be as equally infused with analysis and critique as it is with production, refusing to hide behind modernist claims of objectivity and universal knowledge. Teachers help to construct the dynamics of social power through the experiences they organize and provoke in classrooms.

Critical Science: Responsive Science Teaching in a Political Context

As educators, we often pretend that power relationships do not exist in school settings. From US policies like desegregation to classroom-based activities like group work, we are asked to create the illusion that power relationships are absent or at least negotiable. The story presented at the beginning of the chapter about Cynthia and Jessica reminds us that all children are not equal partners in the process of schooling and that they do not stand at equal levels to construct knowledge in classroom settings.

In the context of science education, assuming equal levels could be devastating as we attempt to infuse our teaching practices with the notion that all knowledge is constructed socially. The question of knowledge construction is not just an epistemological position, it needs to involve serious reflection about identity and experience within the realm of science education as balanced against an individual child's purposes, both within science and the larger culture.

To uncritically accept the knowledge base of science is to perpetuate relationships of power and domination (Barton 1997; Barton and Osborne 1995). Current reform initiatives—even those aimed at enabling students to negotiate their way into the culture of science—if not concerned with helping students critically examine science, may contribute to students choosing to remain outsiders to science and to the culture of science by posing either/or choices about ways of knowing and, hence, identity. Despite the doors that a social constructivist position on subject matter opens (Atwater 1996; Eisenhart, Finkel, and Marion 1996), the otherness created by alternative cultural ways of knowing remains unexplored in traditional pedagogy (Barton 1997; Brickhouse 1994; Stanley and Brickhouse 1995). The borders of science that need to be traversed in school settings require that students who have been marginal to dominant cultures take a step towards involvement. This involves acknowledging the unequal power relations of "knowledge" and "authority" in science (Gore 1993; Foucault 1980) and requires that they willingly examine other culturally defined identities and ways of knowing, but it does not mean that they do so uncritically (Barton and Osborne 1995).

As the stories which will unfold in this book indicate, science education ought to be about more than passing on the disciplinary knowledge of science. Science with children is an incredibly complex social site marked by multiple interacting layers of power arrangements and social and institutional forces which shape and define the boundaries of what is possible. This suggests just how much science education is about issues of power and relationships on a personal as well as an institutional level. As Weiler claims, "teaching extends beyond subject matter knowledge; the centrality of teaching lies in a recognition of the values of students' own voices, subjective experiences of power and oppression, and the worth of their class and ethnic cultures" (Weiler 1988, 148). Envisioning science as a social construction locates the learning and doing of science within social relationships on a day-to-day level, not just historically. Allowing these relationships and the questions and concerns they provoke to guide the science creates opportunities for self- and social empowerment (McLaren 1989).

In short, the essence of a critical science as responsive teaching in a political context lies in a desire and ability to value a multiplicity of cultural experiences, values, and expectations, and then to use those experiences, values, and expectations to challenge status quo beliefs and practices. For example, by valuing the life experiences of all children, teachers and students, in the struggle to create a more inclusive and critical science, can begin to stand up to the institutions of science and education by creating spaces from which to make explicit and problematic cultural biases in the teaching and learning of science (Barton 1998). Children do not engage in science in ways devoid of culture. All bring their histories, values, beliefs, and emotional, social selves to science. As teachers we cannot, deny this or ignore it. In classrooms, our acts and theirs effectively rewrite science. Understanding science as constructed through discourse, as a set of knowledges, that can act as an expression of identity, provides us with the means to reconstruct science and science education so that girls and minorities can find a place or create a new place for themselves within it. Such goals, however, are in conflict with the constraints and demands of an entrenched, established discipline and become the terrain in which science becomes remade. The children's engagement in our science takes on the form of destabilizing activities, challenging and breaking/remaking the grand narrative of science. This makes it into a science for all, or at least a closer approximation.

Conclusions

In *The Post-modern Condition,* Jean-François Lyotard suggests two contradictory societal roles for knowledge, roles which in turn support conflicting models of social structuring: One can decide that the principle role of knowledge is as an indispensable element in the functioning of society, and can act in accordance with that decision, only if one has already decided that society is a great machine. Conversely, one can count on its critical function, and orient its development and distribution in that direction, only after it has been decided that society does not form an integrated whole, but remains haunted by a principle of opposition. (Lyotard 1984)

Such views presuppose a functional rather than passive role of knowledge in society—either actively contributing towards its homogeneity and continuity or playing a fundamental role in its fragmentation and reconstruction. These ideas of knowledge also suggest an active role for knowledge in the creation of power differentials in society, although the outcomes for such differences in each model are radically different.

Working with girls, minorities, and children in poverty is not easy in general, and trying to construct science with these children, science that connects with their lives and empowers them in a liberating manner, has compound difficulties. These children's lives and needs are complex. The manner in which the children in this book explore science allows spaces to fit science into this entanglement rather than keeping it unconnected and separated as so much traditional schooling and science would do. Race, class, and gender, however, are not only dimensions of our social structure dimensions that reflect forms of power and privilege, they are also ways to think about our social processes and the way we live our lives. We argue that relationships are not always smooth or of an actors' choosing; they are constructed in contexts where actors have certain access to power and resources depending not only upon their relationships in the educational structure, but also upon their location and identity in the larger society. As we attempt to consider the multiple layers that emerge and are formed in educational settings, this focus becomes important for understanding the complex problems and issues that emerge in connection with pedagogy and curriculum in classrooms.

The destabilizing acts of the children in the stories to follow in this book remind us of our questions about science, pedagogy, and curriculum; they return us to our question of what to do about difference? For example, How is difference constructed and why does difference matter? When we ask those questions we would like to examine a more foundational one: *Where do things that are mainstream, norms, come from?* When we ask that question, we are looking at both our expectations about "normal" behavior and our beliefs about what is mainstream knowledge in a particular discipline. Why should we take it for granted that a child should learn particular concepts in science or that he or she should even want to? We need to ask those questions if we want to make claims about constructing a science for all.

In looking at the stories to come, pedagogically, in the construction of science curriculum and in our relationships with children, we want to recognize that these are linked and that all must be altered before the marginalized can engage in science. Indeed, our concepts of the role of disciplinary knowledge in society must alter to incorporate change, evolutionary and revolutionary. Underlying our discussion of approaches to dealing with diversity is a questioning of what those courses of action imply about attitudes towards difference. Do they imply that difference is something to be fixed, changed, or that difference should be worked within and maybe finally respected or even advocated? Don't they suggest that difference is fundamentally at the root of the democratic processes of our society?

The stories in this book cause us to think about what doing something about diversity means to us and to society. Doesn't respecting diversity (rather than trying to fix it for instance) imply that our assumptions about the norms of society will change? Doesn't this imply that our ideas about subject matter and assumptions about good behavior, homelife, interests and goals will evolve and enlarge?

We suggest in answer to such concerns a rethinking of science the discipline in trying to construct a "science for all." We echo the classic writing of Peggy McIntosh (1983) in saying that we can't describe what such a new science might be. As our stories suggest it must be emergent through the acts of its creation: "The postmodern would be that which, in the modern, puts forward the unpresentable in presentation itself; that which denies itself the solace of good forms, the consensus of a taste which would make it possible to share nostalgia for the unattainable; that which searches for new presentations, not in order to enjoy them but in order to impart a stronger sense of the unpresentable" (Lyotard 1984, 81).

References

American Association for the Advancement of Science (AAAS). 1989. *Science for all Americans.* Washington, DC: AAAS.

———. 1993. *Benchmarks for scientific literacy.* New York: Oxford University Press.

American Association of University Women. 1992. *How schools shortchange girls.* Washington, DC: National Education Association.

Anderson, C. W. 1991. Policy implications of research on science teaching and teachers' knowledge. In M. M. Kennedy (Ed.), *Teaching academic subjects to diverse learners* (pp. 5–30). New York: Teachers College Press.

Anyon, J. 1984. Intersections of gender and class: Accommodations and resistance by working-class and affluent females to contradictory sex role ideologies. *Journal of Education* 166 (1): 25–48.

Apple, M. 1979. *Ideology and curriculum.* London: Routledge.

———. 1992. The text and cultural politics. *Educational Researcher,* 21 (7): 4–11.

Atwater, M. 1996. Social constructivism: Infusion into the multicultural science education research agenda. *Journal of Research in Science Teaching* 33: 821–38.

Bacon, F. 1963. *The works of Francis Bacon.* James Spedding, Robert Ellis, and Douglas Heath (ed.) New York: Hurd and Houghton.

Ball, D.L. 1993. Halves pieces and twoths: Constructing representational contexts in teaching fractions. In T.P. Carpenter (Ed.) *Rational numbers.* (pp. 157–195). New York: Lawrence Erlbaum.

Barton, A. C. 1997. Liberatory science education: Weaving connections between feminist theory and science education. *Curriculum Inquiry,* 27 (2) : 141–163.

———. 1998. *Feminist science education.* New York: Teachers College Press.

Barton, A. C. and M. D. Osborne.1995. Science for all Americans? Science education reform and Mexican-Americans, *The High School Journal.* 78 (4): 244–252.

———. 1999. Re-examining lived experiences: Radical constructivism and gender. *Journal of Cybernetics and Human Knowing, 6(1)*:47–59.

Bijker, W. 1987. The social construction of Bakelite: toward a theory of invention. In W. Bijker, T. P. Hughs, and T.J. Pinch (eds.) *The Social construction of technological systems: New directions in the sociology and history of technology* (pp. 159–187). Cambridge: MIT Press.

Brickhouse, N. 1994. Bringing in the outsiders: reshaping the sciences of the future. *Curriculum Studies* 26: 401–416.

Buber, M. 1970. *I and thou*. New York: Scribner.

Casey, S. 1997. *Women Invent! Two centuries of discoveries that have shaped our world*. Chicago: Chicago Review Press.

Calabrese, R. and A. Barton. 1995. Mexican-American male students and Anglo female teacher: Victims of the policies of assimilation. *The High School Journal,* 78 (2): 115–123.

Cochran-Smith, M., and S. L. Lytle. 1993. *Inside/outside: teacher research and knowledge.* New York: Teachers College Press.

Cohen, C. 1996. *New ways of making babies*. Indiana: Indiana University Press.

D'Angelo, F. 1987. Foreword. In K. LeFevre, *Invention as a social act.* Carbondale, IL: Southern Illinois University Press.

Davidson, A. L. 1994. Border curricula and the construction of identity: implications for multicultural theorists. *Qualitative Studies in Education,* 7 (4): 335–349.

Delpit, L. D. 1988. The silenced dialogue: Power and pedagogy in educating other people's children. *Harvard Educational Review, 58* (3): 280–298.

Dewey, J. 1902/1956. *The child and the curriculum.* Chicago: University of Chicago Press.

Eisenhart, M., E. Finkel, and S. Marion. 1996. Creating the conditions for scientific literacy: A re-examination. *American Education Research Journal* 33: 261–295.

Epstein, S. 1997. *Impure Science*. Berkeley, CA: University of CA Press.

Fine, M. 1993. Passions, politics and power: Feminist research possibilities. In *Disruptive voices: The possibilities of feminist research*. Ann Arbor: The University of Michigan Press.

Foley, D. 1991. Reconsidering anthropological explanations of school failure. *Anthropology and Education Quarterly*, 22 (1): 60–86.

Foucault, M. 1978. *The history of sexuality*. New York : Pantheon Book.

———. 1975. What is an author? *Partisan Review*, 42: 603–614.

———. 1980. *Power/Knowledge: Selected interviews and other writings. 1972–1977*. New York: Pantheon Books.

———. 1982. *The archeology of knowledge and the discourse of language*. Trans. A. M. Sheridan Smith. New York: Pantheon.

Freire, P. 1971. *Pedagogy of the oppressed*. New York: Continuum.

Gates, H. L. 1992. *Loose canons*. New York: Oxford University Press.

Geertz, C. 1973. *The interpretation of cultures; selected essays*. New York: Basic Books.

Gill, D. and L. Levidow. 1989. Science curriculum innovation and Holland Park School. *Anti-Racist Science Teaching*. London: Free Association Books.

Gilligan, C. 1982. *In a different voice: psychological theory and women's development*. Cambridge MA: Harvard University Press.

Gitlin, A. 1994. *Power and method: Political activism and educational research*. New York: Routledge.

Goals 2000: Educate America Act. 1994. Pub. L. No. 103–227 (33/31/94), Stat. 108.

Gore, J. 1993. *The struggle for pedagogies: Critical and feminist discourse as regimes of truth*. New York: Routledge.

Gorman, M. and W. B. Carlson. 1990. Interpreting invention as a cognitive process: The case of Alexander Graham Bell, Thomas Edison, and the Telephone. *Science, Technology and Human Values*, 15 (2): 131–154.

Guery, F. (1996). The meaning and value of invention. *Science in Context* 9(1), 17–38.

Haraway, D. 1997. *Modest witness@second-millennium. Femaleman-meets-oncomouse: feminisms and technoscience*. New York: Routledge.

Harding, S. 1986. *The science question in feminism*. Ithaca: Cornell University Press.

Harding, S. 1989. Is there a feministy method? In N. Tuana (Ed.), *Science and feminism*. pp. 17–32. Bloomington: Indiana University Press.

———. 1991. *Whose science? Whose knowledge? Thinking from women's lives*. Ithaca: Cornell University Press.

———. 1998. *Is science multicultural? Postcolonialisms, feminisms, and epistemologies*. Bloomington, IN: Indiana University Press.

Helms, J. 1998. Science and/in the community: context and goals in practical work. *International Journal of Science Education* 20(6), 643–653.

Hollingsworth, S. 1994. *Teacher research and urban literacy education*. New York: Teachers College Press.

Hooks, B. 1994. *Teaching to transgress: Education as a practice of freedom*. London: Routledge.

Hubbard, R. 1990. *The politics of women's biology*. New Brunswick, NJ: Rutgers University Press.

Keller, E. 1985. *Reflections on gender and science*. New Haven, CT: Yale University Press.

Kirkup, J. and L. Keller. 1992. *Inventing women: science, technology, and gender*. Cambridge, UK: Open University Press.

Ladson-Billings, G. 1994. *Dream keepers: Successful teachers of African-American children*. San Francisco, CA: Jossey-Bass.

Lave, J. 1988. *Cognition and practice: Mind, mathematics, and culture in everyday life*. New York. Cambridge University Press.

Lee, O. and S. Fradd. 1998. Science for all, including students from non-english language backgrounds. *Educational Researcher*, 27(4): 12–21.

LeFevre, K. 1987. *Invention as a social act*. Carbondale, IL: Southern Illinois University Press.

Lensmire, T. 1994. *When children write: Critical re-visions of the writing workshop.* New York: Teachers College Press.

Longino, H. 1990. *Science as social knowledge: Values and objectivity in scientific inquiry.* Princeton: Princeton University Press.

Lubar, S. 1987. Culture and technological design in the 19th century pin industry: John Howe and the Howe Manufacturing company. *Technology and culture* 28: 253–82.

Lyotard, J.F. 1984. *The postmodern condition: A report on knowledge.* Trans. G. Bennington and B. Massumi. Minneapolis: University of Minnesota Press.

MacIntosh, P. 1983. *Interactive phases of curricular re-vision: a feminist perspective.* Wellesley MA: Center for Research on Women.

Maher, F. and M. Tetreault. 1993. Doing feminist ethnography: Lessons from a feminist classroom. *Qualitative Studies in Education,* 6 (1): 19–32.

Maher. F.A. 1987. Inquiry teaching and feminist pedagogy. *Social Education,* 51:186–192.

McCarthy, C. and Crichlow, W. 1993. *Race, identity, and representation in education.* New York: Routledge.

McGinn, M.,K., and W. M. Roth. 1999. Preparing students for competent scientific practice: Implications of recent research in science and technology studies. *Educational Researcher,* 28: 14–24.

McLaren, P. 1989. *Life in schools: An introduction to critical pedagogy in the Foundations of Education.* White Plains, NY: Longman.

National Research Council (NRC). 1996. *National science education standards* Washington, DC: National Academy Press.

Noffke, S. 1995. *Educational action research.* New York: Teachers College Press.

Osborne, M. 1997. Balancing individual and group: A dilemma for constructivist teachers. *Journal of Curriculum Studies* 29(2): 183–194.

Reyes, P. and R. Valencia. 1993. Educational policy and the growing Latino student population: problems and prospects. *Hispanic Journal of Behavioral Sciences* 15(2): 258–283.

Rodriguez, A. J. 1998. Strategies for counterresistance: Toward sociotransformative constructivism and learning to teach science for diversity and for understanding. *Journal of Research in Science Teaching*, 35 (6): 589–622.

Rosebery, A. 1994. Interpretive perspectives on scientific discourse. Paper presented at the American Educational Research Association, annual meeting New Orleans, LA.

Rosebery, A., B. Warren, and F. Conant. 1992. Appropriating scientific discourse. *Journal of Learning Sciences*, 2: 61–94.

Rosser, S. 1990. *Female friendly science*. New York: Teachers College Press.

Roth, W. M. and M. K. McGinn. 1999. >uuDELETE science educations/lives/work/voices. *Journal of research in science teaching* 35(4): 399–422.

Roychoudhury, A. D. Tippins, D., and S. Nichols. 1995. Gender-inclusive science teaching: A feminist constructive perspective. *Journal of research in science teaching* 32(9): 897–930.

Stanley, W. and N. Brickhouse. 1995. Multiculturalism, universalism and science education. *Science Education* 78: 387–398.

Vygotsky, L. 1978. *Mind in society: the development of higher psychological processes*. Cambridge MA: Harvard University Press.

Weber, M. 1949. *The methodology of the social sciences*. Glencoe (IL) Free Press.

Weiler, K. 1988. *Women teaching for change*. South Hadley, MA: Bergen and Garvey Publishers.

West, C. 1993. *Race matters*. Boston: Beacon Press.

Yates, J. and A. Ortiz. 1991. Professional development needs of teachers who serve exceptional language minorities in today's schools. *Teacher Education and Special Education* 4: 11–18.

Young, R. 1989. Racist society, Racist science. In Gill and Levidow (Eds.) *Anti-Racist Science Teaching*. London: Free Association Books.

PART ONE
ELEMENTARY SCIENCE EDUCATION AND SCIENCE TEACHER EDUCATION

Chapter 2

The Magical and the Real in Science and in Teaching: Joy and the Paradox of Control

Margery D. Osborne
David J. Brady

Tale of a Paradox (Narrated by David)

Lately, I work with particularly simple physical systems. I work with optical interferometers, which to me are the most beautiful of instruments. Interferometers are magical to me, they represent what I love about science. If I have been a bad scientist, spending too much time writing grant proposals and going to committee meetings, interferometers act as nthe jealous mistress and will not work for me. If I have been the good scientist, I can quickly make them work, but there is little science in this—it is by feel, practice, intuition. The interferometers that I love form stark beautiful fringes from white light. To do this, two mirrors separated by a few centimeters must be aligned to within one tenth of a hair. This alignment is achieved by turning a micrometer. With the proper alignment, fringes appear. The fringes are not hard, dangerous, and stable like a laser beam, however. They are playful and dynamic, shimmering, moving, bouncing about as one hovers, breathing nearby.

Aligning an interferometer can take hours. One nudges knobs, cleans surfaces, and squints at spots. During this time one is immersed in a world, lost in that world, suspended. This is an existence outside of experience for it is without self-consciousness. To do this work, one is an artisan, an artist, and a human, but to name these states implies self-consciousness.

Joy, the emotional, spiritual, transformative state, is found through experiences in which multiple ways of knowing and ways of experiencing (i.e., intellectual *plus* emotional, spiritual *plus* sensual) occur. This is sometimes an easy combination or sometimes a conflicted intersection, but both create metaphorical places in which the uncontrolled, the unpredicted, the dreamed of are created. For us a component of joy is in this sense of the uncontrolled (loss of self), but we also assert that joy is achieved through control (self-control). It involves a willful belief in the unbelievable (such as a child's belief in Santa or the magical world of Borges or Coelho, in which people are choosing to live within an alternative logic). It is achieved through both a consciously and unconsciously chosen limiting of our control of the environment. Joy is magical, strange, and subversive, but we strive to achieve it and we, as adults, do so by choosing to act without control. We choose to not be smart or competent in the domains that would allow us to have control. We ask though, as we reflect on such experiences in the context of teaching and of science, How do we get there (so that we can act in such a way) and what are the ramifications?

In this essay we would like to write about joy and about magic. The stories we will recount of our work in science (like David's), or with practicing teachers or small children, are examples of magic; all are mysterious, transformative. And we know that all have rational, scientific, professional explanations, but as scientists and as teachers we are content to leave them as magic as unexplained and uncontrolled. We recognize that fully rational examination would kill our understanding, as discussion kills humor, or that rigorous analysis of magical understanding would take so much time, space, or money that other important aspects would be lost. We focus on magic because the word is provocative and we wish to provoke an exploration of a much neglected facet of science and of education, the uncontrolled and out of control, the qualities of science and of teaching that cause joy. We do this because the joy that we derive from these acts is what draws us in, reinforcing and shaping our immersion in both enterprises. It drives our engagement in both.

We present this in counterpoint to both utilitarian and critical schools of philosophy/sociology of science and of science education. There is a lot to be said for thinking about whimsy, playfulness, passion, spirituality, beauty, and many other words rarely applied to science or education. Thinking about these informs our understandings of what we do and why and could probably and profitably shape our conceptualizations of what we ought to do. To construct an argument for such an assertion, we will present a number of short narratives of our work with children and teachers, narratives which illustrate qualities of magic. There are three important qualities of magic that are present in all our stories. These are

- seeing things not seen before
- seeing things as something else from normal
- seeing things that aren't there at all.

All of these arise from seeing old and everyday things in new ways. This is done by

- using things inappropriately
- looking at things too hard
- being in a different head space
- doing things the wrong way
- playing with not understanding.

These are the qualities that constitute our construct magic, but in themselves they are not magical. What they do however is create

- parallel worlds
- coexistent and incommensurate interpretations
- coexistent meanings that are orthogonal.

These also are not magical. What makes these magic is the lack of control we can exert over them and the feelings these arouse in us, and we assert these feelings are those of joy.

The stories we recount to illuminate qualities of the joy that we find through magical acts in teaching and science come from three sources. David talks about his work as a practicing physicist and electrical engineer at the Beckman Institute, University of Illinois, and Margery narrates stories from her graduate teaching in science education with practicing teachers and her work with "at-risk" preschool children. The three contexts are intertwined; Margery and David collaborate in the construction of the graduate class and Margery brings to bear her experiences as a research scientist as she works with small children and tries to make sense of David. We begin by attempting to paint a more substantial picture of the places from which we draw our stories, and then we will move on to the stories themselves. We will interlayer our stories with the development of the construct magic in the teaching and in the doing of science. Through the stories we make an argument that irrespective of our best (or worst) efforts to the contrary, children carve out spaces of agency in which they make (or remake) their worlds. It is a conceit to say that the teacher does this for them. What a teacher should be doing is "seeing" and enabling the world they create and understanding it in the sensitive way an empathetic adult can. We call this an education in the tradition of magical realism (Ellsworth 1997), and we draw on the work of Jorge Luis Borges and Paulo Coelho to describe it. We also draw from the philosophical writings of Ludwig Wittgenstein and Martin Heidegger to argue the political ramifications of such a theory as well as the humanist traditions of Robert Coles and Bruno Betelheim. This culminates in our creating an argument for a new appreciation of joy, in science and in teaching.

The Context In Which Margery Works

I have been working for the past two years as a volunteer in a preschool classroom. This is in addition to my work as professor at the University of Illinois. At the University of Illinois I teach graduate classes in science education; I work with full-time graduate students (those doing a thesis) and practicing teachers. In my graduate classes I try to challenge students to think hard about the nature of science and how our ideas about what constitutes science shape our teaching. My work with preschoolers informs this as I use this setting and my work there to examine my own beliefs about teaching and subject matter. In the preschool classroom I work the morning session three days a week. Last year I tended the "water" table, this year Molly, the classroom teacher, has asked me to supervise "boxes and junk," an imaginative construction project for children. Children must qualify as at-risk to be in this classroom. This means that they must meet at least three criteria from a list of a dozen possibilities, including being a child of a single parent family, having low income, having poor language development, being born prematurely or with low birth weight. There are fifteen children in the room, a lead teacher, an aide, and me.

The Context In Which David Works

I design digital optical imaging systems and I teach graduate and undergraduate classes in imaging system analysis and design. I expect visual representations to become natural components of technologically enabled human communication, and my research group is working to make interfaces between image collection systems and human use as smooth as possible. That is how my collaboration with Margery began; it seemed to me that a test bed for such integration might be science education. My group has prepared digital imaging kits including compact cameras and transmitters for use in elementary science education. Under the direction of Margery and her graduate students, children have used these kits to explore the insides of their throats, the lives of dogs and cats, science experiments under water, the insides of their parents' furnace, what can be seen from a kite, etc. Our goal is to discover new approaches to integrating automated and human data fusion. I believe that integrated, automated, and human-directed systems allow more people to learn and to express themselves faster. Margery and I are also exploring this concept in her graduate classes with the "Cognoscence" web interface that I wrote. Cognoscence is a hyperlinked bulletin board with integrated text and image posting. The idea is to allow individuals to tag and interrelate visual information in real time.

The Magical and the Real: Poetics of the Everyday

We extend Wittgenstein's thoughts on the poetry of ideas to our thinking about objects (Perloff 1996; Wittgenstein 1983). The poetry of ideas, when applied to objects, is what occurs when we look at an ordinary thing for so long it becomes strange. When this happens we begin to see new qualities in it and inappropriate uses for it. This is a both a perverse act and a subversive one. It is possible because all things can act as metaphors for something else. All things have overlapping parts or qualities. It is always plausible to say something is like something else (the ocean is like a cloud, a rainbow is like a fish). These metaphors are both plausible and provocative for while in some senses the ocean *is* like a cloud, in other senses it is not. The metaphor invites comparison and critique as well as enabling new understandings by pointing out likeness. As applied to objects, how is a toilet paper roll like a wheel? How is it not? When we look at toilet paper rolls do we see wheels? What enables a child to see such things?

James Makes a Fox from Dried Leaves
Molly is working with James. James is four. They are reading the book *Look What I Did with a Leaf!* and discussing the animal shapes that the author has made out of dried and pressed leaves (Sohi 1993). How did she do these? What shapes are there in leaves? What structures and textures? How did she arrange them? James points to the fox—he wants

Figure 2.1 Jame's fox

to make it, too. The shape of the body is an oval. He selects an oval leaf. The fox has a backbone, so does the leaf. The leaves are too rounded for legs, so he cuts them very carefully in half. Laying them out and creating the pattern is hard for little fingers. Gluing the leaves down requires patience. James writes his name when he is finished, and the word "fox" (see fig. 2.1).

Reducing the fox into geometric shapes and then seeing similar shapes in the leaves is an abstraction; as a process it involves abstracting a generalizable ideal and fitting it to a pattern. Patterns are substanceless descriptions of relationships. Patterns, both seeing and making them, are compelling in science and in general because they cause us to see the totality of a phenomenon in new ways. The parts of the phenomenon that don't fit the pattern both become invisible and are thrown into relief. Pattern is compelling because the act of bringing order to disorder is infused with romantic mystery and power. But the parts that are left in disorder are even more mysterious and maintain the phenomenon's own power.

In what ways are leaves like foxes? How is a whole composed of parts? In creating his fox James engages such questions. The leaves, as James handles and examines them, assert their own qualities. This can happen because they possess both generalizable qualities and irregularities: those qualities that exist and which we can't classify using the particular set of patterns we are trying to impose. The imposition of theories—generalizations, patterns—enables seeing the phenomenon in new ways because of the abstracting qualities of the process and also because this process is situated in a creative act, in this case trying to make something out of something else. When the qualities of the phenomenon that don't fit the pattern begin to demand our attention, stop being invisible, the assessment of the object or the pattern or the task will be revised.

It's because pattern exists as an overlay on the surface of the real phenomenon that irregularity and regularity can coexist—the imposition of patterns or generalizations doesn't change the object's true nature. There is more to the phenomenon than can be described by the pattern. Recognition of that causes a child or teacher or scientist to apply existing patterns to new phenomena and to discover new patterns. The person is the creative agent in the dialectic between pattern and irregularity which intersect within the phenomenon. We assert *our* reality through the imposition of patterns, the recognition of regularities, the creation of explanations. The dialectic between the person and the phenomenon, the pattern and the irregularities, the explained and the unexplained drives the inquiry.

There are two different ways that patterns are looked at and used. We can look at the pattern and the object through the pattern, using the pattern to give us new ways to see that object in order to continue contemplating the object. The pattern is a tool to enable seeing the object itself. Or the pattern is a tool for doing something with the object. When James composes his fox from the leaves, he is engaging this second process. The richness of his picture, though, comes from the irregularities, the imperfections. The poetry of ideas, when applied to James and his fox, describes the process he has gone through to see a fox in a leaf, to see a composition such as he has created out of its unlikely parts. The leaves are strange and so is the fox. Certain qualities and meanings of leaves and foxes are what is being negotiated here, and this occurs through their inappropriate use. Addressing the question of how a child is able to see such things and engage such comparisons suggests to us that we need to consider both the qualities of the child and of the teacher/teaching. The same process of imposing simplifying patterns on complex and multidimensional phenomena describes the act of teaching. That process allows us as teachers to do certain things, including seeing in a new light the qualities of a child, qualities which amaze us.

42 The Magical and the Real in Science and Teaching

Scott and Darius: Boxes and Junk
(with Margery Narrating)

Little boys, once they reach a certain age, see cars, trucks, motorcycles, wheeled vehicles that go *fast* everywhere. I have been working, for about ten years, enabling children as they design and construct things from their imagination. Sometimes we design things for a specific task (making bubble structures for example (Osborne 1999)), sometimes just for the pleasure of making things. Lately I have been guiding young children in a construction activity called "boxes and junk." Below is a picture of Scott holding one of his creations. I worked with Scott last year, and this year when I stopped by his classroom he showed me his skateboard (see fig.2.2).

He constructed it from cardboard, toilet paper rolls, paper, and masking tape. The rolls of paper extending off the back are "the *wind,* because it goes so fast!"

This is a truck made by Darius (see fig.2.3). Darius is four and in the at-risk preschool classroom where I am working. It is made of a Styrofoam egg carton, a molded plastic cookie tray, cardboard cylinders, and glue. He has painted it teal with tinted glue. He mixed the color. The whole thing took him three full days to make, the first time he has ever made anything or sustained an activity for more than a few minutes running (according to Molly, the classroom teacher). Since making this truck he has made two more, of big molded Styrofoam shapes used to pack

Figure 2.2 Scott holding his skateboard

Figure 2.3 Darius's truck

computers and of Legos. Molly hung his first truck on the wall so all the children could admire it. Before he would let her do this he wrote his name on it. He's never done that before either, Molly claims. Every day he asks if he can take the truck home to show his mama. In my two years in the classroom this is the first time I have heard him speak a full sentence.

****

When we consider how a child becomes able to do such things and the teacher's role in this process, we begin to meditate on ideas similar to Donald Schon's (1984)observations on the generative qualities of metaphoric thinking in professional problem-setting. Schon describes metaphor as creating a perspective or way of looking at things, and the application of a new metaphor enables new ways of seeing, hence his phrase "generative metaphor." It is possible for these metaphors to coexist both harmoniously and in conflict, and for this reason they also serve to engender inquiry into objects, actions, and the problem itself. Like Wittgenstein suggests in his writings on language and meaning, when we look at a truck or a skateboard assembled from old boxes, we see something absurd, but in articulating the absurdity we bump against our understandings of trucks and trash and children. By looking so closely at these things that they become absurd, they loose their everyday meanings and we rediscover (to our amazement) the magic within them. This poetry of ideas, applied to objects and actions, becomes the site of discovery of new meanings and functions as a heightened form of social and cultural critique.

The children's creations amaze and amuse us. They do cause us to think new ways about the possibilities in tras,h but when we talk about "social and cultural critique," that occurs as we look at the children. When we see the possibilities in allowing them the space to be creative, they challenge our assumptions about what they are capable of doing. As teachers, the site of critique and creativity is there, in seeing new potentials in the children, in seeing the children in ways we haven't imagined them to be. We bump against our ideas of the child and also our ideas of what constitutes teaching for here teaching, is an act of discovery. It is surprising.

The "bumps" of Wittgenstein have much in common with the "breaking" of Martin Heidegger (Dreyfus 1991; Winograd and Flores 1987). According to Dreyfus (1991), Heidegger suggests that we act in this world without contemplation and that detached contemplation, while illuminating, is also unnatural and obscures phenomena by isolating and categorizing them separately from the contexts in which they occur. Currently our beliefs about rational, scientific thought and processes as well as our conceptualizations of best practice in teaching idolize such idealizations of the thought-action connection. According to Winograd and Flores (1987), "Heidegger does not disregard this kind of thinking, but puts it into a context of cognition as praxis—as concernful acting in the world. He is concerned with our condition of throwness—the condition of understanding in which our actions find some resonance or effectiveness in the world (1987, 32–33)." We act and know how to act because of this quality of throwness or immersion in the everyday. We become aware of the qualities of objects and contexts because the effectiveness of our unreflected-upon actions break down—for some reason they do not work. Heidegger terms this "breaking." "[O]bjects and properties come into existence when there is a breakdown in that [structural] coupling" (Winograd and Flores 1987, 72). In working with a child like Darius, of course we think we know what we are doing, and we do, otherwise he would not have responded as he did, but still the qualities of his response surprise us. They are and are not what we predicted. As teacher, we are working on a knife edge of discovery. It would be possible to fall off to either side. We are gambling, taking a chance. The payoff when we win is huge.

But in talking about the poetics of the real or the metaphors inherent in the acts of children's construction or of teaching we are making sense of what children and teachers do, but we are not really making much sense of why they do these things. What causes people to put themselves in settings in which things become strange? To address this question we turn to magical realism (see D'haen 1995). Magical realism as a term describing a literary genre was appropriated from the world of art criticism. There it referred to that artistic form characterized by realistic depiction of the paradoxical and strange. The effect of such work is both unsettling and faintly shocking. Examples would be Salvador Dali's paintings of clocks and Rene Magritte's depictions of the incongruously assembled living room. Such early work in the genre evolved into Marcel Duchamp's urinals, *Bicycle Wheel*, and other ready-made art, or recent "difficult art" (Yenawine 1996) such as the recent "Sensation" exhibit at the Brooklyn Museum of Art. In recognizing that, we also recognize the connection to Wittgenstein's poetics of ideas, for that is exactly what Duchamp's art concerns. And although there seems to be little that is poetic in Chris Ofili's "Holy Virgin Mary," it certainly achieves its purpose in making us uncomfortable.

Heather's Example

Heather is working three slide projectors simultaneously; over the lens of one she has placed a red film, over the next green, over the third blue. She moves the red, green, and blue lights so that they lie on top of each other. The light that results is white. She shows us an image she has created from a photograph taken in Thailand. It is made by separating the colors, red, green, blue, from her picture and creating separate negatives, first in red, then green, then blue light. Recombined, the colored images are transformed: cyan, magenta, yellow. The final image, returned to black and white, emerges through a textured pool of light.

Heather shows us that when she casts a shadow in red light with her hands, the shadow that we see is cyan; in green light the shadow is magenta. The students are amazed at her explanation of their origin; they cannot seem to understand, Why are the combinations not *combinations*? How can white light be colored light? The students go to the slide projectors and use their hands to create magical roosters, swans, leaping rabbits in a rainbow of colors. Their shadow theater evokes the *wayang kulit*—the traditional puppet shadow theater of Java and much of Southeast Asia (Thai: *nang talung*). Such performances of the shadow puppet plays are interpreted in religious or mystical fashion. In them the ancient animistic rituals of Southeast Asian religions are played out. As in the shadow theater, the translucent screen becomes heaven, the floor becomes earth, and the puppets are man, the seeker.

Figure 2.4 Thailand photograph

****

According to Teresa D'Haen, magical realism manifests fundamental qualities of postmodernism. It exhibits or engenders, "self-reflexiveness, metafiction, eclecticism, redundancy, multiplicity, discontinuity, intertextuality, parody, the dissolution of character and narrative instance, the erasure of boundaries, and the destabilization of the reader" (D'Haen 1995, 192). Both D'Haen and Elizabeth Ellsworth (1997) point out that magical realist text speaks from the margins and takes the reader empathetically to that place. It does this through the device of realism; the writing, through its realism, draws the reader into a magical place where the laws of Western logic and rationality are suspended. The reader is seduced into thinking from a perspective other than their normal one. The reader moves into an alternative world from which they can critically reflect back on the realities of the normal world. Heather, in her pictures created through the application of science, incites mystical, empathetic understandings of another culture as well as the mysteries of science. This sense of mystery is itself a challenge of the rational beliefs we hold concerning what science understandings should be. Making the pictures involved understanding the technology of the printing technique; understanding the pictures involves an opening of the mind to mysteries.

Telling Stories at the Water Table
(Narrated by Margery)

Conversations at the water table are wonderful. Today the water table is full of cornmeal. Sometimes it has rice, macaroni, mixed beans, snow, brown sugar, even water in it, but today it is cornmeal. In many ways cornmeal is my favorite. It has a beautiful color and smell. It is texturally complex—gritty but smooth—and cool to the touch. It compacts, so it can be molded and used to build structures. It accumulates a static electrical charge which causes individual grains to leap and jump or form interesting swirling patterns at the bottom of table. It is quite fascinating! Children come to play with it and can engage in exploring its properties for a long time without tiring of it. They can also use it as a component of imaginative fantasy play such as cooking, building houses and landscapes. As we play, conversations arise spontaneously while children discuss their activities and discoveries.

I am playing myself; after setting up for the morning I sat down and began by trying to use a bulb baster to suck up cornmeal and blow it back out again. If done carefully this creates a puff of particles. I didn't discover this, Darius did. K'nisha and Tiffany are pouring cornmeal from one container to another using a funnel. I send them back to get smocks on, and when they return, they settle into this activity—pouring cornmeal from

Figure 2.5 Two girls playing with cornmeal

small jars and bowls into large containers, filling up the large containers, and then pouring those into the basin of the table with much satisfaction. Tiffany grabs a tube and begins stirring the cornmeal as she adds it to a jar. She is talking about making soup and K'nisha chimes in. I ask them, "What kind of soup do you like?" and this begins a lengthy conversation about the likes and dislikes of everyone and anyone. Conversations with four-year-olds are patchworks of things they do and know, things they have seen, and basic fantasy. I wonder about Tiffany—does she make supper for her baby sister (which is what she is telling me)? I know she did take care of the baby until her mother, an alcoholic, abandoned them. Then her aunt had custody until she, too, abandoned the children. K'nisha on the other hand is a mystery to me. She comes to school with her hair freshly put into twisty ponytails, carefully finished with *Goody*™ barrettes. Her clothes are clean and she is, too. The criteria for qualification for at-risk services are manyfold.

While this conversation is going on, Andrew and Jake arrive. K'nisha and Tiffany continue their fantasy cooking projects, but Andrew and Jake have begun molding a building with the cornmeal. This causes a certain

Figure 2.6 Boy playing with cornmeal

Figure 2.7 Girl playing with cornmeal

amount of competition between the two groups, for space and for cornmeal, and actually the boys start stealing cornmeal from the girls, causing a great deal of outrage and some whining. Not too much though—both girls are pretty assertive. Finally the girls start building castles, too. Theirs involve piling the jars and bowls on top of each other and then sifting the cornmeal down the sides. Occasionally they will mound some cornmeal on the tops and mold this, then brush it off again.

The conversation turns to home renovation. Jake, like Tiffany, is also four years old and Molly calls him the primary caregiver for his baby sister. His mother is what I'd term an opportunistic sex worker at a local bar managed by her mother. The children have different fathers and the family lives in the local trailer park. Jake is into construction. The talk at the table is a complex discussion of construction principles and electrical wiring. How do you wire a castle made of cornmeal? Interesting question I'd say, and finally I do. I ask, "What kind of lights are you planning on installing?" Jake says, "Christmas tree with colors." I ask, "Where'd you learn how to do that wiring?" Andrew says, "Second shift with the good old boys."

****

Borges' (1962) story, *Tlon, Uqbar, Orbis Tertious*, is an example of magical realism, which effects many parallels to the stories the children create as they play. In Borges' story magical realism is constituted from a juxtaposition of real and imaginary worlds. It is created in such a way to enable many and layered interpretations, to cause reflections on things taken for granted as well as to construct alternative pathways to sense making. Borges plays in his essay in between the real and fantasy worlds of fact and fiction, and the imaginative world is portrayed as reality. When we read his work we enter his fantasy and, in doing so, rethink our own identity, for identity is fundamentally shaped by our assumptions as we make sense of the world. These assumptions are disrupted and questioned, and the children do the same as they engage in their fantasy play.

According to one tradition in child psychology we are all, and especially children, constructing meaning in our lives and of our lives (Bettelheim 1975). Bettleheim argued for the primacy of narrative fantasy in this process: "For a story to truly hold a child's attention, it must entertain him and arouse his curiosity. But to enrich his life, it must stimulate his imagination; help him to develop his intellect and to clarify his emotions; be attuned to his anxieties and aspirations; give full recognition to his difficulties, while at the same time suggesting solutions to the problems which perturb him" (Bettleheim 1975, 5).

Fantasy is not a form of escapism but is a way of working through the problematic nature of life without being defeated by it. Meaning is constructed by overcoming the odds and transcending the immediate problematics of existence. While Bettelheim is talking of fantasy worlds found in fairy tales we would argue that children, as they engage in play such as we have been describing in these stories, are creating fairy ,tales of their own and through these fairy tales they construct and reconstruct the world in which they exist. We are reminded of the great lesson learned by the shepherd boy in Paulo Coelho's *The Alchemist* : "[A]t a certain point in our lives, we loose control of what's happening to us, and our lives become controlled by fate. That's the world's greatest lie" (1994, 20). The story goes on to describe how the shepherd boy pursues his "personal legend," learning to become in control of his own fate.

We argue that if we pay attention to children's worlds—the "real" and those that they construct—we should be able to recognize a play of realities and meanings. The children's activities at the water table are a magical mix of fact and fiction, a reconstruction of truth, a reconstruction which enables their lives. They use the social context of this play to work out important relationships and understandings about words such as caring and friendship. The text is subversive and "encourages resistance to monologic political and cultural structures" (Zamora and Faris 1995, 6); they do not create the lives that white middle-class adults would wish to impose on them. In this sense, it can be said that the magic in their stories "is often given as a cultural corrective, requiring readers to scrutinize accepted realistic conventions of causality, materiality, motivation" (Zamora and Faris 1995, 3). Through the magic in their fantasy play the children engage the difficulties in their lives. We would argue that although the difficulties of these children seem extreme, difficulties are unavoidable—they are an intrinsic part of human existence—but if we are not afraid, recognize and meet these difficulties, we can reach an understanding that is *enabling*; and the children in their play are reconstructing reality to do this always and in every setting, including "science time" at the "water table." Finally, we as teachers, existing as we do in privileged places, have no right to be afraid when confronted by children engaged in this process, and indeed if there is joy to be found in teaching it is through enabling this and savoring it when it happens.

Final Story: A Child's Magic

Andrew is three years old in this story. He lives with his father, who is in his forties and works two jobs. He also lives with an older sister who is seven. When his father is working the night shift, Andrew sleeps at his grandmother's home. His mother never sees him although she is in the area. He and his sister were taken from her care by the Department of Family Services. The preschool program here is configured as a family program with four home visits a year by Molly, the classroom teacher. At home his environment is rather restricted; Molly describes it as "Spartan," clean and very, very neat. His activities there center around the computer, playing games of any type for all hours. At school he would choose to do this also. There are two computers in the classroom and Andrew would gladly monopolize both simultaneously, running back and forth between them clicking buttons with ferocious intensity and without waiting to see what happens.

At this moment Andrew is running back and forth across the room. He pauses for a second by my side, "Look, Margery, this is my shapeshifter. "He holds up two plastic counters; a square and a triangle attached. He pushes it at my face, holding it like a gun.

52   The Magical and the Real in Science and Teaching

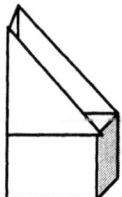

Figure 2.8 Shape shifters

Margery: Oh yeah, how's that work? (I wince and jerk my head away, thinking he's going to make ray gun noises and talk magical superhero powers.)
Andrew: Like this. (He holds it up again and rotates it four times around an imaginary, stationary center.)
Margery: Oh! What's that called?
Andrew: An octagon.

****

We began this essay claiming that magic has three components:

- seeing things not seen before
- seeing things as something other than "normal"
- seeing things that aren't there at all.

Andrew is doing all these things and he is using things inappropriately, looking at things too hard and doing things the wrong way. He is also out of control. However, the teacher is the one who is in a different head space. The magic here is both in the transformations Andrew enacts with the counters and in the transformations Margery must enact to align herself with his thinking. Anthony Greeley in *Ecstasy: A Way of Knowing*, describes such transformations as mystical, as a means of coming to know a greater truth: "In summary, then, there seems to be an evident convergence among the various attempts to describe the mystical experience. It is a breaking away from everyday life and an instantaneous, fantastically powerful immersion into a transformed unity which illuminates the person, exalts him, and transforms him, at least temporarily. He sees things *the way they are* and finds himself in the possession of a power much greater than he, which overwhelms him with joy" (1974, 24). In this essay we have argued that the magic of science and teaching is created through a poetic understanding, through the poetics of the everyday and of ideas. Gaston Bachelard tells us, in *On Poetic Imagination and Reverie*, that understanding is reached through synthesis of the "whole" and the whole includes the magical, the absurd, and the impossible:

When a poet tells us the of secret of milk he is not lying, not to himself or to others. On the contrary, he is finding an extraordinary totality. As Jean Paul Sartre says, "*we must invent the heart of things if we wish one day to discover it. Audiberti informs us about milk in speaking of its* secret blackness. *But for Jules Renard, milk is hopelessly white, since* it is only what it appears to be."

Here we can grasp the difference between the dialectics of reason, which poses contradictions in order to cover the entire range of possibilities, and the dialectics of imagination, which would seize all that is real, and finds more reality in what is hidden than in what is visible. (italics in original, 1971, 8–9)

We called joy an emotional, spiritual, transformative state at the start of this essay and claimed that it was achieved through an intersection of multiple ways of knowing—the synthesis that Bachelard describes. We claim it is achieved as a dialectic between control and lack of control. It involves an awakening, a surprise. In many ways the uncontrolled is how the other ways of knowing come into play and what they might be (emotional, spiritual, etc.), for we assume we are acting within the rational/intellectual and then the other surprises us. We are surprised because our assumptions about how to act or what is going on are challenged. Because our reality is challenged, we arrive at a state of critique, and it is here that creativity happens. Creativity is fundamentally about critique, and that is what joy arises from.

There are qualities in the people involved in these stories. They exhibit self-absorption, tunnel vision, an intensity in their playfulness as well as great courage. There is joy in what our students are doing in these stories—the joy of creativity and discovery, of amazement. There is also joy for us as teachers. All of this is contained though, shaped by constraints of purpose, setting, and materials. For us a component of joy is in this sense of breaking through these controls and achieving the uncontrolled. It is achieved through both a consciously and unconsciously chosen limiting of our control of the environment. Joy is magical, strange, and subversive, but we strive to achieve it and we do so by choosing to act without control. The products of such actions can be amazing.

The concerns of education, schools, teachers seem to be all about control. Our preservice teachers loose sleep over the question of how to achieve it. The practicing teachers we work with in developing inquiry approaches to science teaching worry about test scores and covering the curriculum—questions deeply rooted in issues of control. David, as a practicing engineer, is faced daily with creating particular things for particular audiences *on demand*. Margery is heard to mutter that she wishes people would do what she tells them. We suggest, however, that maybe much is to be gained from celebrating our successes when we are not in control, when the unexpected happens. Choosing to advocate children or teaching or even science out-of-control, ecstatic, makes us uncomfortable however. Such a discomfort seems more than one purely about teaching and the purposes of education. In writing this essay we talked about this discomfort at length. David wrote this in an email to Margery:

I acknowledge the place of ecstacy in my life and the role of joy in my science, however, . . . I have dark thoughts. Reading the January '00 *Harpers*, I find the comments of a hereditary peer, Viscount Monckton of Brenchley, in favor of "action against cruelty to animals, particularly fishing with rods. All cats to be muzzled outside to stop the agonizing torture of mice and small birds." I hate catch and release fishing, which seems like fish torture to me. But I am uncomfortable with outlawing it. I can accept the violent dismemberment of mice for the joy of cats. Are mice necessary to ecstasy in cats? These words make me uncomfortable. I am in favor of pleasure but against torture. I favor joy but not ecstasy. Words have meaning and yet . . . I am progressive, not puritanical. I am more than "progressive," a word for a cautious liberal or a rambunctious conservative. I love the wild, uncontrolled, dangerous, joyous, ecstatic.I've been listening to Fred Eaglesmith's (1995) song "Wilder Than Her." "She's a summer storm, I'm a hurricane. One blows through town, one blows town away . . . When we go driving in our cars, racing through the night she can drive as fast as me but she stops at all the lights." I recently bought a four seat convertible. This is the paradox, the paradox between joy (the convertible) and self-control (4 seats, stop lights). Learning and teaching. How can one as wild as me teach, how can one less wild ?

There is something ineffable about the concept of joy. It seems that it is good if we recognize it in retrospect, but if we pursue it maybe not. We return to Greeley's writings on ecstatic ways of knowing; he is discussing ways of knowing beyond the rational and logical:

The fourth form of knowledge, the mystical, resembles the metaphysical in that it is concerned with the ultimate, is like the mythopoetic in that it does not deal with logical propositions or the laws of discursive reasoning, and is like science in that it comes into contact with the hard data of external reality. It is different, however, in a number of critically important ways. It is daemonic (in Rolo May's sense) in that it takes possession of the whole personality far more than any other knowledge. It is immediate. One comes into contact with the real without the need for either prose or poetry; it requires neither logical proposition nor symbolic representation. The real is encountered directly and as it is. . . . [It] cannot be expressed adequately in symbols because it is obtained without them—co-naturally, as it were, through the immediate union of the knowing subject and the object known. (1974, 60)

There is the quality of this "immediate union of the knowing subject and the object known" in all the stories we tell. We describe a form of experience involving more than the intellect, one which involves the whole person. In reflecting on our experiences with this we are tempted to argue that pursuing such a way of knowing drives us in our science and teaching, and maybe in stating this we are arguing for the recognition that this drive articulates something basic about the human condition. In calling this union "joy" we are describing the emotional state we feel when we achieve this union. This reflects a tension, a paradox: there is a passive aspect in this as well as an active one, for we need to create the conditions under which joy occurs but then relax our control so that it can happen. Joy reflects a position of power, power over and acquiescence to power. It is an acknowledgment of power and *that* is magic.

References

Bachelard, G. 1971. *On poetic imagination and reverie.* (Trans.) C. Gaudin. Dallas, TX: Spring Publications.

Bettelheim, B. 1975. *The uses ofeEnchantment: The meaning and importance of fairytTales.* New York: Vintage.

Borges, J. L. 1962. Tlon, Uqbar, Orbis Tertious. In *Labyrinths: Selected stories and other writings.* New York: New Directions Book, 3–18.

Coelho, P. 1994. *The alchemist.* San Fransisco: HarperCollins.

Coles, R. 1968. *Children of crisis: A study of courage and fear.* New York, Dell.

D'haen, T. L. 1995. Magical realism and postmodernism: Decentering privileged centers. In *Magical Realism: Theory, History, Community*, (eds.) L.P. Zamora, W.B. Faris. Durham NC: Duke University Press.

Dreyfus, H.L. 1991. *Being-in-the-world: A commentary on Heidegger's Being and time, Division I.* Cambridge: MIT Press.

Eaglesmith, F. 1995. Wilder than her, *Drive-in movie.* Poughkeepsie, NY: Vertical Records. ASIN: B000007MW5.

Ellsworth, E. 1997. *Teaching positions: Difference pedagogy and the power of address.* New York: Teachers College Press.

Greeley, A. M. 1974. *Ecstacy: A way of knowing.* Englewood Cliffs, NJ: Prentice-Hall.

Osborne, M.D. 1999. *Constructing knowledge in the elementary school science classroom: Teachers and students.* New York: Falmer.

Perloff, M. 1996. *Wittgenstein's ladder: Poetic language and the strangeness of the ordinary.* Chicago: University of Chicago Press.

Schon, D. A. 1984. Generative metaphor: A perspective on problem-setting in social policy. In *Metaphor and Thought*, ed. A. Ortony. Cambridge: Cambridge University Press.

Sohi, M.E. 1993. *Look What I Did with a Leaf!* New York: Walker.

Winograd, T. and Flores, F. 1987. *Understanding computers and cognition: A new foundation for design.* New York: Addison-Wesley Publishing.

Wittgenstein, L. 1983. *Philosophical investigations.* Oxford: Blackwell.

Yenawine, P. 1996. Difficult viewing: Unsafe subjects in recent American art. Miller Committee Lecture, University of Illinois at Urbana-Champaign.

Zamora, L. P. and Faris, W. B. 1995. *Magical realism: Theory, history, community.* Durham, NC: Duke University Press.

Chapter 3

A Room of One's Own: Concrete and Conceptual Spaces

Heidi Bulmahn Barker

> Not a flat, not an apartment in back. Not a man's house. Not a daddy's. A house all my own. With my porch and my pillow, my pretty purple petunias. My books and my stories. My two shoes waiting beside the bed. Nobody to shake a stick at. Nobody's garbage to pick up after.
> Only a house quiet as snow, a space for myself to go, clean as paper before the poem.
> —Cisneros, *The House on Mango Street*

A Room of One's Own

Jennie's classroom is her own. It is warm and inviting. The students' desks are arranged in five groups of four and this cluster forms the focus of the classroom. It is evident that this is where the central activity of the class takes place. There are no chalk boards and the teacher's desk is unassumingly placed at the back of the room. Children's work hangs on the walls—both individual projects and classroom ideas. There is a table in one corner that is piled with examples of leaves and other plant life that the children have brought in. Laminated projects from former third-grade students also sit on the back table as evidence of the family atmosphere and value in student work that Jennie holds in her room. Along another side of the room is a class set of plants growing under fluorescent grow lights and placed on a self-watering system. The Wisconsin Fast-Plants are just beginning to form flowers and several of the students look at them before they sit down in their seats. They have just returned from physical education. On this particular day, there are five adults in the room. There is a special education assistant, a grade-level teaching assistant, myself, the principal, who is evaluating Jennie's teaching, and Jennie. It doesn't seem crowded and the kids welcome the adults into their projects willingly, as does Jennie. She is comfortable in her room and in her teaching and makes all of us feel a part of the teaching and learning that occurs. Her warm and inviting classroom is a reflection of the teacher and teaching that has created this space.

From the moment that preservice teachers begin their formal education as education majors they speak about having their own classroom. I know this from listening to the elementary education students whom I teach and also because I once was one of these teachers waiting to create my own space for teaching. Preservice teachers watch and occupy the space of mentor teachers. They take all of the differing ideas that they see and make their own ideal for what their teaching will look like. They see more than the decor of these rooms. They see attitudes and ideas that they want to replicate and those that they want to disown. They crave the opportunity to design and implement their own ideas. A teacher's classroom is his or her own space. It is a space that is occupied not only physically but also mentally while engaging in the creative act of teaching and making decisions about the whats and the hows of the ways that the curriculum will be enacted. Many factors influence these decisions (colleagues, state, national, and local standards, administration, community, etc.), but it is the different ways that every teacher creates his or her own space for teaching that is interesting to me. Teachers create their own spaces for making changes in pedagogy and curriculum. The metaphor of a room of one's own, based on feminist writings on space, place, and relationships, describes this process.

Virginia Woolf wrote about creative process (and I believe teaching is a creative process) as a "room of one's own." As Woolf describes a room of one's own where women writers need to have space and freedom to create, teachers also need to have the freedom and space to think about teaching practices and make these their own. This approach to curriculum and pedagogy emphasizes the construction of understanding of reform efforts and is based on the belief that there is no single, effective method for teaching and learning as students and teachers differ in their preferences, desires, imaginations, attitudes, and interests. As the teachers "do" science activities, try activities out with their students and in their classroom settings, and reflect on these experiences, their knowledge of science and science teaching is expanded. In order to make this new knowledge a part of their understanding of teaching practice, each individual must create an understanding that meshes with their prior science teaching experiences. Like a blank paper before writing a poem

(Cisneros 1984), there is much possibility for ideas, but expectations, attitudes, emotions, imaginations, and desires play a part in how the poem is created.

In this chapter, I will further elaborate on this metaphor, using narratives from my current research project to illustrate how teachers create spaces for understanding and implementing reform efforts in their own classrooms. The project that I have been working on tries to set up spaces for teachers to make their own understandings of curricular and pedagogical changes in science teaching and learning. I will describe the project, tell stories of two teachers involved in the project, and describe part of their journey in making changes in their classrooms. By telling their stories, I hope to relate that when teachers make decisions about the creative process of teaching and are given space do this creating, they form their own understandings of curricular and pedagogical reforms.

Teachers as Space Makers

My current research project is working with a group of elementary teachers in a small district who are designing and implementing new science curricula. The district is moving away from a text-oriented science program to an inquiry, process-based approach that includes hands- on activities, integration of other subject areas, and in-depth study of topics. The funding for the project is from a state science literacy grant and will last two years.

The textbook-purchasing rotation made it the year for new science textbooks to be selected in the elementary buildings. One of the teachers had gone to a curriculum development workshop and science camp with educational consultant and former science coordinator for Champaign Schools, Marilyn Sinclair. This event became the impetus for an Illinois State Board of Education (ISBE) Science Literacy Grant written for the school district by Marilyn Sinclair and Margery Osborne, my academic advisor. The teachers on a district-level science curriculum committee with representatives from each grade level decided to make a commitment to moving away from a set of textbooks in each classroom to a process-oriented approach to teaching and learning science. The grant was written to support the teachers in this commitment. Part of the school improvement plan was also written in conjunction to this commitment. "The district will continue to develop an inquiry based approach to science instruction through a grant from the ISBE" (Balmoral Community School District, 1998).

The grant was written as a two-year proposal and was a cooperative effort between the school district and the Department of Curriculum and Instruction and the University of Illinois at Urbana-Champaign. Through this collaboration, the school district will gain "ongoing consultation and guidance in science curriculum design to meet state and national standards and the provision of graduate-level classes designed specifically for the Balmoral teachers and their needs in science teaching, including: 1) enhanced knowledge of science concepts and principles, and 2) instruction and experience in inquiry methods of teaching" (ISBE 1999, 1). The graduate course is taught in the summer in conjunction with a summer science camp for elementary students in the district. Additional staff development, planning meetings, and support continue throughout the school year with a cooperative effort between the teachers, the administrators, and the university consultant and graduate student.

After initial meetings that included teachers on the science curriculum committee,[1] the curriculum coordinator for the district (who is also a grade school principal), a university professor (Margery Osborne), and the science consultant (Marilyn Sinclair),[2] the teachers made it clear that they needed support "in their need for increased knowledge in the content of the sciences appropriate for elementary classrooms, and of the pedagogical skills for teaching an inquiry-based science program" (ISBE 1999, 4). The way that the grant is set up allows for the teachers to learn content and pedagogy in the same manner that inquiry learning for children proposes (Easley 1990). The teachers try out the teaching methods, ask questions, discuss their observations and experiences, and look for answers. Time for conversations between teachers, between teachers and university personnel, and between administrators, teachers, and university personnel is integral to this process and is written into the grant by providing substitutes, planning time, and in-service time. The teachers are given support to try out the methods not only during the summer science camp, where reflection on teaching and learning also occurs, but also in their classrooms during a two-year pilot period. During this pilot period the teachers try out the units they have chosen and developed, discuss with one another their experiences with the teaching and learning, and modify the unit for future teaching.[3] The ideas and questions of teachers are valued in the process of change (see Spillane 1999; Schwille et al. 1983) and are used to guide the process of implementing the new curriculum.

When teachers are given spaces (both conceptually and concretely) to make their own sense and understanding of their emotions and attitudes about science and science teaching, they are given a powerful place[4]. It is the power to create understanding for themselves, making the new a part of their own knowledge and creating a way for new methods to become their own. When teachers work together to support one another in their learning efforts, in trying new ideas, and in talking about and reflecting on their ideas and practice, they gain a certain strength and self-confidence in the ways they work with each other, their students, and their ideas. In time, they develop a broader understanding of learning, teaching, and education, an increased knowledge of themselves as learners, and a more complex understanding of the nature of science. Both factors, having individual spaces and spaces to share community conversations, seem to be important in their understandings and in developing a critical reflective way of approaching science teaching and learning.

Community Conversations

"Collaborative conversation" (Hollingsworth 1994) as a method of inquiry may challenge the traditional ideas of research. How can simply talking together about concerns of practice be a method of research and how can talking together as a means of support contribute to the knowledge of both the teacher and the researcher? To me, this seems a very logical, practical, and realistic way of creating spaces for change. Conversation allows for a supportive structure and a format for gaining understanding. The playing field is leveled as the teachers become experts by talking about their practice, and the power of the researcher is taken away as he or she is not the expert, but a part of the conversation. The relationships formed through the conversations create a different context for talking that is not a "dialogue—similar to the conversation in a play or novel, which appears to have two or more voices, but which actually comes from one author's perspective"; it is more than "simply a discussion of prearranged topics and readings through a formal discourse structure" (Hollingsworth 1994, 8). It is a collaborative and sustained conversation, an exchange and reformulation of ideas, intimate talk, and reconstructive questions. Throughout the conversation, common stories about learning to teach are understood. The "vague, almost subconscious questioning and tentative knowing about teaching school [is] thus elevated, voiced, and connected" (Hollingsworth 1994, 8). The conversations that teachers share specifically about their teaching experiences give those experiences meaning for the teller and for the listener. The teller is an expert and the teaching experience is important enough to take time to listen and learn from the story. When teachers talked about their experiences with the new science curriculum, reflected on their practices and planned for future teaching, the teachers articulated their own understanding of the process. This articulation began to validate and connect their experiences to the ways that they view science teaching and learning.

Collaborative conversations were enacted in various forms as teachers began to implement the inquiry-based science curriculum. They took place between teachers of the same grade, between teachers across grade levels, between teachers and administrators, and between teachers and myself. As I visited classrooms, attended planning meetings, and participated in the summer science camp, we discussed stories of teaching in the context of exchanging ideas, questioning ideas, and how teaching and/or learning experiences were connected to one another. These conversations became my research "data." As teachers formed a closer relationship with the material and with each other, the conversation continued and contributed to reformulating their ideas and questions about their science teaching. The ideas and questions about their science teaching were what they began to explore more closely in the space of their own classrooms.

Alice's Story: Creating a Space of Her Own

Alice has been teaching for more than twenty years. She is a second-grade teacher who did not actively participate in the science reform efforts during the first year of the grant. By not actively participating, I mean that she did not pilot any of the second-grade units and she was not a member of the district science committee. Alice did participate in grade-level planning meetings where teachers met to talk about, look at, and evaluate possible curriculum, topics, and materials. She became aware of the curricular and pedagogical changes that the district had made a commitment to through the ISBE Grant. In the spring of 1999, Alice signed up to teach a class during the district's summer science camp. She "wanted to find out more about the new science program." By participating, Alice would also gain four hours of graduate credit as the first two weeks of the camp are a graduate class for the teachers where they are given an opportunity to learn about inquiry methods of learning by doing science themselves. Part of the requirement of the course was to keep a journal documenting these experiences. Alice wrote:

"I know the first day we met I wasn't sure what to do with the roly-poly and sow bugs. Those were things I would just as soon step on, but it only took a few minutes and I was noticing all kinds of things about them.

Then we were asked to draw and write about what we saw. My first thought was I don't want to do this, but then it became fun and I learned in spite of myself."

Alice was one of the teachers who volunteered to share her writing with the teacher group. She had written a poem about the bugs.

> Mr. Bug, I have a question
> Are you a roly poly or a sow bug?
> Whenever we're asked we say
> See if you can make us curl into a ball
> and roll around on the ground.
> Mr. Bug, I have a question
> Are you an insect?
> Whenever we're asked we say
> Count our legs to see if we have more than six.
> Mr. Bug, I have a question
> Where is your habitat?
> Whenever we're asked we say
> Look under old wood, flower pots, or crocks
> Wherever it is damp or wet.

Alice told me that she had forgotten how much fun it was to observe and look at things. She was a teacher who taught from the science textbook, making it through the entire book every year. Again from her journal:

> In the past the class has read the chapters together in Science and answered the questions at the end in a 30 minute block of time. Next year I hope to arrange my schedule to have a larger block of time, do more hands-on science in groups of 2 or 4. We will do more chart work and journal writing. Also, I will be doing more linking of curriculum. I hope to work with the art teacher to learn some of the art techniques we might use during science activities. I want to make these changes because science camp has shown me that children respond to this kind of learning with eagerness and curiosity. Not only do they learn information, they also gain social skills and responsibility. There are more opportunities for problem solving and thinking skills to develop. Participation in summer science camp has provided me the opportunity to participate and see this first hand.

I show these examples from Alice's journal because she has changed her view of teaching and learning science so radically and attributes the change to her firsthand experience and opportunity to try out the ideas herself. Alice found the "space" that allowed her make the changes. She needed to see and experience inquiry science teaching and learning for herself before she could understand and envision the way that it would work in her own classroom. This experience gave her the confidence to try the changes she proposed in her summer journal, and her experience in the classroom has encouraged her to continue.

I have visited Alice's classroom several times this year, and each time she has greeted me smiling warmly and eager to show me the projects on which the children are working. Her room is filled with books about the topic of study, and the walls outside her classroom are covered with artwork and writing related to the topic. Her students are sitting in groups of two. She told me last week that she is glad that the district has changed to this curriculum. "It is more interesting for me and the students. They really look forward to science time and the activities. They enjoy the learning."

Recently, the second-grade teachers met to talk about the curriculum, share their experiences with the units that they have been teaching, and talk about problems, concerns, and suggestions they have about the new science program. This school year is the first that all of the second-grade teachers are teaching the new curriculum. They all taught a unit on butterflies at the first of the year and Alice was eager to share her successful experience with "failure."

> My caterpillars from the company was a terrible experience. They arrived late, they did everything that we were supposed to study over the weekend, so our lessons on observing the changes were pretty much nonlessons. So I had an idea. I quickly wrote a note home asking everyone to bring in a caterpillar that they found, being careful to note where they found it and what it was eating. I didn't know how many I would get. It was the end of the season, getting cold and all. I had so many different caterpillars! We learned so much—what they all ate, what kinds of different changes they all went through, and on and on. The kids just kept them on their desks and brought in what they ate. It really turned out great and all because of a mistake in the ordering! One of the kid's caterpillar came out of its cocoon right in the middle of math—we just all stopped and gathered around his desk. He was like a proud father with his new butterfly!

Alice's excitement in her story was passed to the other teachers. She was the expert as they asked her question after question about how she structured the activity. How much time did she allow for the lessons? How did the kids do all day with containers of caterpillars on their desks? How did she make time for these extra activities in the curriculum? To answer these questions, she said fairly confidently, "I just don't worry about it too much. I figure we'll just keep studying until we're done." Alice, who said that she remembered how much fun it was to learn when she did the science activities herself, was seeing this excitement in her students. Because of her own experience, the doing of the science activities and the motivation that these activities provided for continued learning became the center of her science curriculum. She let that experience influence the way that her science teaching looked and the way that her classroom looked and felt as she taught and talked about science with her students.

Alice was "given" a space to work through her own perceptions of what science teaching and learning are. She had time in the science camp to try the new pedagogy and theories for herself and with students. There was a structured time to talk about this experience and an assignment to write about her experiences. Teachers supported one another as they team taught the summer classes. When Alice returned to school in the fall, this structured support continued as her building principal continued to support the science program, encouraging teachers to implement the program in their classrooms. I believe these given spaces did contribute to supporting the pedagogical and curricular changes that Alice implemented in her classroom. But what I believe to be more important is the space that Alice "created" for herself. She made changes in her own perceptions of what science teaching and learning are. She gave herself permission to understand what it meant to be interested and excited about learning and she began to implement these new perceptions into her teaching. Without the space that Alice created for herself, the imposed structures to facilitate change could make no difference. Alice, in her own space and in her own classroom, holds the power for the success or failure of school reform efforts. Structures designed to implement pedagogical and curricular changes need to acknowledge, facilitate, and celebrate the importance of that power.

David's Story: Conversation for Making His Own Understanding

David was not an original member of the science committee. Because of his own interest in science teaching and curriculum changes that would affect him, his teaching, and his students, David became involved in the project from the very beginning. He did not participate in the summer science camp, but he did come to the first meeting of the district science committee in the fall of 1998. The meeting took place in one of the fifth-grade classrooms. We met here because this particular teacher used tables for student work places and it lent itself well for a meeting of about fifteen adults. There were representatives from each grade level as well as Margery, Marilyn, the administrator in charge, and myself. This was the first time that I met any of the teachers. It was also an introduction to the structure and wording of the grant that would fund the process of changing their science curriculum. This exchange took place between David and other teachers at that meeting.

David: So, there are only four unit topics to be covered the entire year? Do you see them lasting a full nine weeks each?
Jennie: Well, there is so much material and the kids are loving it. I'm hoping that I can get it all done before I have to move on.
John: That's the mentality that we have to get away from. Now it's time for Science. Now it's time for Social Studies. We need to start thinking of the learning as a whole.
David: That's easy for you to say, but I switch with another teacher and am locked into a schedule. I suppose we could combine classes and I teach science every day for a half hour. Will there be enough to last a whole nine weeks?
Liz: It's a different approach. Instead of a smattering of subjects, we are going in depth on a few.
David: So it's like if you teach multiplication in third grade then you don't get it again until it comes up in high school? Once the subjects are taught they aren't covered again.
Liz: But now we are teaching things that aren't appropriate for some grade levels. This scope and sequence looks at the grade levels and it isn't the only time a subject is taught.

The teachers that answered David had already been piloting some of the units in their classrooms or had been a part of the summer science camp where they had tried some of the teaching. David had had no experience with the new curriculum, but was concerned about the way that it would impact the way his classroom and the science curriculum he had been teaching worked. He wondered what it meant to go in-depth, to only study a few topics, and how this would impact his teaching and his students. The new program directly challenged the way that he perceived his success in teaching. In order for him to accept the change he would have to understand how a process-approach science curriculum fit into his understanding of teaching and learning.

As I got to know David, through visits to his room, through grade-level planning meetings, and through informal conversations, the reasons why David asked and continued to ask these questions became more clear to me, and perhaps to him as well. David really wanted to understand the underlying reasons for the curricular and pedagogical changes and he wanted these reasons to make sense to him. David took teaching seriously. He, like the other male teachers at the school, wore a tie every day. He was the only member of his immediate family and much of his extended family to go to college and to leave the small blue-collar town in which he grew up. Education was important to him. David lives in the community, his own children go and will go to school here, and his wife is also a trained teacher who works as a teacher assistant in the school. He volunteered to pilot the unit his grade level was piloting to "better understand the changes being made."

One example of this occurred as soon as I entered his third grade classroom about halfway through the grant cycle. The students were out to recess and David was straightening things up, getting ready for the science lesson that would take place upon their return in a few minutes. "Hi, David! How are things going?" I asked as I came in.

David smiled and said, "I want you to get this on tape." He went on to tell me that he thought there was no way that the new science curriculum was going to work, how there was no way that the students could engage in the inquiry for a full nine weeks, that the school district's test scores were going to fall—which is the way that the community, the administrators, and the teachers judge their work.

> Number one, I'm scared for the test scores. I mean maybe some others are happy being on a third grade level, but we've always been on a fifth grade level. We have done well. We have so much to teach and I've already spent four weeks on this and it is doing, but I don't know what the kids are taking away. We have so much to cover. I've got the incubator, the experiments I promised the kids we could do and its not fun for me or for anybody when we're rushed. They don't know the vocabulary. I feel that they need a foundation before they do the hands on stuff for it to make any sense to them. And, in the last week they've gotten tired of it. I think today's lesson will be good because they get to compare the different rocks and minerals, but don't know what to do with myself. I don't feel like I'm teaching when the students are doing. They like it, but are they learning? I've tried to do the kit exactly as it is presented to really see what it is like and I feel like there needs to be a balance.

It is evident in the way that David talked about the new curriculum that he has many concerns that need to be addressed if the new curriculum is going to play any role in his science classes. He wanted to make sure that his goals for teaching met the expectations of his community, the state, his administrators, and most importantly that they delivered what he believed to be good science teaching. He wanted to "know" that his students came away from his classroom ready to go to fourth grade, ready to pass the standardized tests, and ready to understand some basic science facts and concepts. At this point, David was not convinced that the new curriculum would meet these expectations.

David continued to ask questions and began to make adjustments in his teaching to fulfill his own need to fit the new curriculum into his own understanding of his role as science teacher. "I am definitely more of a guide now. I've added a few vocabulary words that we've read about and we are going to have the students identify the rocks and minerals by name because I feel that the parents want them to be accountable and I do, too! Without these things it is hard to evaluate the students. How do you evaluate thinking skills and how much hands-on is tested on the achievement tests?"

David was also concerned about the bigger picture of the topics taught and the depth of the topics taught across the grade levels. Part of the reason for a change in curriculum and the discussion of the science curriculum, according to David, was a lack of communication between grade levels and within grade levels about what science was taught. Each teacher "did their own thing" and some topics were covered every year and others were never covered. David now sees this as not such an important issue. "If the focus is on the thinking skills and the processes, it really doesn't matter what the topics are." This was stated as part of a discussion of teachers being "territorial" about their science unit topics. Teachers are very protective of the spaces they create. David had a wider view of this space as he worked to understand the impact of the new curriculum on the district as a whole. He asked the high school science teacher if he thought that the elementary curriculum would prepare the students for high school science classes. The high school teacher responded in a positive manner. For David, the answer became part of his own understanding of where his teaching fit into the district's science program. For him, the understanding of the program and its ability to meet the expectations of everyone David is required to please, including himself, is the search that David continues.

Virginia Woolf gave women the advice to write for themselves: "[I]t is much more important to be oneself than anything else. Do not dream of influencing other people. . . . Think of things in themselves" (1957, 111). Perhaps this is the advice that David should take. Teach for yourself; teach what and how feels right to you. This is the dilemma he faces (and with which many teachers struggle). How can you work for your community, with your colleagues and administration, and with your students and also teach for yourself? How can all of these ideas negotiate a space in teachers' classrooms?

Questioning the Process

These questions of what teachers teach and how teachers teach are not a new debate. What children learn and the ways that teachers set up learning in their classrooms are topics that have been discussed for decades in our public schools. Dewey asks, "Is the experience a personal thing of such nature as inherently to stimulate and direct observation of the connections involved, and to lead to inference and its testing? Or is it imposed from without, and is the pupil's problem simply to meet the external requirement?" (1944, 155). From where should the stimulus for learning come? What is the role of the teacher? How does learning become meaningful for children? The meaning is made in the way the subject matter is experienced by the learner (Dewey 1944; Duckworth 1987). By making children familiar with their world in such a way as to "capture their interest, to let them raise and answer their own questions, to let them realize what they can do. . .[teachers can give them tools] so that they have the interest, the ability, and the self-confidence to go on by themselves" (Duckworth 1987, 225). To do this, teachers need to feel that they have the freedom to diverge from their plans—to value the questions and insights that they and their students make along the way of their teaching and learning. "If teachers feel that their class must do things just as the book says, and that their excellence as a teacher depends upon that, they cannot possibly accept children's divergence and children's creations" (Duckworth 1987, 225). How can children be taught to think, to question, and to problem solve, if the goals and objectives of the lessons are so specifically stated that they seem to stifle creative teaching and learning? When the teachers ask me, "How will the kids score on the state test?" it is that feeling that the kids must do exactly as the book says in order for the teacher to know if they came away with the lesson goal that leads them to this dissonance.

How will changing the ways that science is taught and learned and the content of the science that is taught and learned affect the state tests? This question came up over and over as I talked to teachers who began to try the curriculum in their classrooms. They seemed to like the enthusiasm of the children for learning science and the involvement of the kids in their projects and learning, but that question lurked in the back of their minds. As the teachers were given more control in the decisions they made about the direction that their curriculum and teaching would go, they began to wonder if these decisions and directions were the "right" ones. Do I believe in this? Is this what the state, the district, my students, the community, and the administration expect and want from learning? The journey of that process is important. How do teachers negotiate their place within changes in pedagogy and curriculum?

Spillane labels this negotiation as "zones of enactment." He contends, as I do, that reform efforts need to include both external supports for teachers in the form of time for "deliberations about the substance of the reforms and the practicing of these reforms" and material resources to support these reform efforts *and* space for these external resources to interact with the teachers' own individual practices (1999, 171).

My role has been to support the classroom teachers in the design and implementation process. I do this by helping during science lessons in classrooms, listening to teachers talk about their experiences with the curriculum, and participating in planning meetings where teachers share with each other their experiences with teaching the new curriculum. In this role, I have seen the teachers try out new teaching practices and try to develop an understanding for them. They want to understand and feel that they are doing the right things, that their students are learning, and that they are teaching. In teaching in a more process-oriented approach, they have given up some of their control and the understanding that they had of teaching and learning. They are taking their new experiences with learning and creating a new understanding of their own place.

David and Alice are examples of teachers who began to make changes in their practice, reforming what teaching and learning look like in their classrooms. Their understandings were created very differently as they worked in their own ways to fit the new curriculum into their previous experiences and understanding of science teaching and learning. David needed to seek answers to his questions. He conversed with other teachers, administrators, and others to work toward an understanding of the benefits and pitfalls of the new curriculum. In this way, his research made him an "expert" on the process and he began to look at the big picture, defending the program as he worked to implement it in his classroom. Alice was convinced of the merits of the changes as she became excited about her own learning and her students' enthusiasm. Through the support of her principal and other teachers in her building, she felt her classroom was a safe place to experiment with the curriculum. She had permission to try the hands-on projects without worry of making mistakes. She felt the learning was in the experience. By looking at their stories, we see how these teachers created rooms of their own. They were given space and freedom to create the ways that the new curriculum would become a part of their own classroom teaching. Their classrooms became a reflection of the teachers and the teachers' own understanding of the reform efforts.

Individual Spaces

The two stories from teachers I work with on this project are examples of different ways that particular teachers have created spaces for making the science curriculum their own. They tried to answer questions about the process of reforming the science curriculum for themselves, but in very different ways. Alice worked towards her own understanding of her role in the process of changing science curriculum by doing the science herself and remembering how exciting learning can be. For David, it was not enough to see his students engaged in the process of learning. He wanted to justify and answer the questions about the curriculum and his role pragmatically. I have chosen these stories not only because they support my concept of individual spaces, but also because they show that these spaces are both concrete and conceptual. The teachers talk about changes in the ways that they think about teaching and learning science and also changes that they have made in their own actions and in doing the teaching and learning of science.

The teachers also make these changes in their practice in ways that point out what is problematic in the metaphor. A room of one's own implies individualistic creation when teaching is anything but; it implies a powerful position when it is not, it implies that they have had the space and the time to think about their practice and apply this to their teaching. Woolf's metaphor in relation to women writers having a space void of the other responsibilities of women's lives is unrealistic—women cannot be free of their other responsibilities such as mothering and societal and household tasks, nor would all women want to be fee of their responsibilities. We learn to balance and juggle these responsibilities, finding the space for them to overlap and influence one another. Women writers do not work in a vacuum where they concentrate only on writing. Women's lives are complicated. Women have had to put together a mosaic of activities and resolve conflicting demands on their time and attention (Bateson 1989, 13). Bateson utilizes the metaphor of a crazy quilt to describe the way that women "compose" the experiences of their lives. The experiences/quilt may look "willy-nilly," but "the various pieces, wherever they come from, have to be trimmed and shaped and arranged how they fit together"(Bateson 1989, 13). I saw the teachers make sense of curricular and pedagogical changes in a similar way. As I listened to their conversations during meetings, as they reflected on "doing" science, on their teaching, their lessons, and their classroom experiences, they wove their experiences, creating conversations that moved from one experience to another but in some way connected to the understandings that they were creating for themselves. As the teachers take in other's perspectives, new ideas, and new experiences, they work to find "the essence of themselves" (Lindbergh 1983). By creating structures (like meeting times and systems for collecting teaching materials) that gave them time to reflect on these experiences, share their ideas with their colleagues, and try these ideas in classrooms, the teachers formulated their own individual understandings of the curricular changes. These changes became part of their individual teaching and the room of their own.

A teacher's classroom is his or her own, but the space is created by a mosaic of influences. Teachers do not work in their own creative vacuum, but they do have influence over the patterns that emerge and over which themes are present in the foreground of their work. These stories show that teachers can and do make decisions about the creative process of teaching, work to make changes in their practice, and make decisions about the ways that these processes are enacted within their classrooms. The way that they implement the change process is a personal one, and because it is a personal process, school reform efforts largely depend on the teachers making these efforts a part of their own thinking, teaching, and learning. By giving teachers a space to understand their conceptions about science learning, teachers learn to articulate their own philosophy of education—what they think about the curricular and pedagogical reforms. As the teachers did the science themselves, reflected on this experience, and then tryied the processes with children, they began to form their own understanding of the curricular and pedagogical changes.

References

Balmoral Community School District. 1998. School report card. Balmoral, IL.

Bateson, M. C. 1989. *Composing a life.* New York: Atlantic Monthly Press.

Cisneros, S. 1984. *The house on mango street.* New York: Vintage Books.

Dewey, J. 1944. *Democracy and education.* 1916. Reprint, New York: Collier Books.

Duckworth, E. 1987. *The having of wonderful ideas and other essays on teaching and learning.* New York: Teachers College, Columbia University.

Easley, J. 1990. A teacher educator's perspective on student's and teacher's schemes. In *Science education: A minds-on approach for the elementary years,* ed. E. Duckworth. Hillsdale, NJ: Erlbaum Associates.

Hawkins, D. 1974. Messing around with science. In *The informed vision: Essays on learning and human nature.* New York: Agathon Press.

Hollingsworth, S. 1994. *Teacher research and urban literacy education.* New York: Teacher's College Press.

Illinois State Board of Education. 1999. Programs for scientific literacy: Staff development projects. Springfield, IL.

Illinois State Board of Education. 2000. Programs for scientific literacy: Staff development projects. Springfield, IL.

Lindbergh, A. M. 1983. *Gift from the sea.* 1955. Reprint, New York: Vintage Books.

Schwille, J., A. Porter, R. Floden, D. Freeman, L. Knappen, T. Kuhs, and W. Schmidt, 1983. Teachers as policy brokers in the content of elementary school mathematics. In *Handbook of teaching and policy,* ed. L. Schulman and G. Sykes, 370–91. New York: Longman.

Spillane, J. P. 1999. External reform initiatives and teacher' efforts to reconstruct their practice. *Journal of Curriculum Studies* 31 (2): 143–75.

Woolf, V. 1957. *A room of one's own.* 1929. Reprint, San Diego, CA: Harcourt Brace Javanovich.

Notes

1. The committee is composed of ten teachers who represent all grade levels, kindergarten through eighth grade.

2. Marilyn, Margery, and I have multiple roles as participants in the grant and the reform efforts. Marilyn Sinclair is a retired science teacher and district-level science coordinator. Her role is mainly leading formal planning meeting, dealing with the administrative aspects of the grant and the district, and helping to set up the systems that keep the materials and supplies available for teachers. She also acts as the administrator of the summer science camp and helps to teach the graduate course. Dr. Margery Osborne is a faculty member in the Department of Curriculum and Instruction at the University of Illinois at Urbana-Champaign. Margery's main role in the grant is teaching the summer graduate course and teaching a class of elementary students enrolled in the summer science camp. She is also an invaluable resource and support as we work through the reform efforts in the classroom. As the graduate assistant, I try to support the teachers' efforts. I visit classrooms (as another set of hands, as a teacher, as a participant in the lesson), I participate in planning sessions and meetings (both informal and formal), and I listen to the teachers and try to act on their needs. I also teach an elementary class during the summer science camp and assist with the teacher course.

3. During the 1998–1999 school year, 10 units were piloted. Fourteen teachers were involved in the teaching of these units. During the 1999–2000 school year, a total of 22 units will have been piloted. All teachers who teach science in the district elementary grades (K-6) are involved in the teaching this year (30 teachers).

4. This implies that teachers will use the space that has been given to them to work toward change. For example, this "space" could be in the form of time for reflection and conversation or in time and support for trying out new ideas. When I say "given," I mean provided by or supported by others who hold power (administration, colleagues, community).

Chapter 4

Teaching in the Interface: Reflections on Urban Science Teaching

Kathleen St. Louis
Tanahia Burkett
Angela Calabrese Barton

Angie: Does the shelter own this land here [behind the shelter]?

Shelter Administrator: No, but you can use it in the after-school science program if you want.

Angie: Yes. I am thinking the children and I might explore the land and figure out what kind of long-term study we could do here. Maybe we could study the animals here over time, or maybe we could build some tree houses or we could plant a garden. In the last shelter where I worked we planted a garden. I want to see, first, what the children are interested in.

Shelter Administrator: You can use the area, but I don't think a long-term project is a good idea. I want you to do "stand-alone" activities each week so that when children move into the shelter or leave the shelter they don't have to also enter and exit any long-term science project. They also like to do activities where they take something home, something that they can call their own.

For us this conversation captures a set of tensions inherent in our work teaching science with homeless children. We want to provide children with the opportunity to explore the world through scientific lenses. We want to help them gain access to the practices, language, and skills of science in authentic and meaningful ways. Yet, we want to be sensitive to the children's lives and the physical and psychological aspects of those lives. But, what does this mean? What does it mean to construct a meaningful and empowering teaching and learning setting across these domains of science and homelessness and poverty? Building such a bridge (or bridges) was part of the motivation behind constructing long-term scientific inquiries with youth. We did not want to do cookbook, forty-five-minute exercises like we often see done in school. We wanted to spend some time learning about the children's questions, interests, and desires and work with them to construct meaningful science from their lives. Yet, as the shelter administrator pointed out, long-term inquiry denies the transient nature of the children's lives. They may never see the endpoint of their work. She also worried that long-term projects meant the children would have fewer products to take home with them. And, her points are well taken. Over the past five years we have been struck by homeless children's desires to use science class to make things that are useful in their lives. In fact, our experiences teaching science with children in poverty have taught us to rethink science education as being about pragmatism and utility just as much as it is about learning the canon or anything else.

We begin with this short vignette because it raises for us in dramatic ways many tensions of teaching science. Specifically, it raises two questions: First, what does it mean to do emergent science with children in urban poverty? By emergent science, we mean science that is driven by students' "life-worlds." Thus it is science that emerges from the intersection of the interests, values, experiences, and beliefs of students and their life-worlds. Second, how is emergent science transformative, and for whom is it (or ought it to be) transformative? In other words, how might being open to the kinds of questions and ideas raised by children in poverty be transformative not only for the children, but for the teachers and science as well?

In the remainder of this chapter, we examine these questions regarding emergent science by using our teaching of science to homeless children as well as critical-feminist theories. To do so, we present a series of postcards. Like postcards sent to friends during vacations and travel, each postcard presented here provides the reader with a snap shot of one (or more) of our teaching dilemmas related to emergent science and transformation as well as our initial attempts to describe the meaning this image has for us. We have specifically chosen mundane examples because we believe that emergent and transformative teaching rests in seeing and responding to ordinary classroom events in different ways. Critical reflection of and response to everyday situations allows experiences, such as the ones represented in the following postcards, to become powerful ones. Although each postcard raises its own question, all of the questions dance around a common thread of emergent science and transformation. Each of our postcards is followed by a discussion critically examining what it means – or might mean—to do emergent science with children.

Postcard 1: Cultural Conflict—Insider or Outsider?

This postcard shows us teaching science at a homeless shelter. We are struck at once not by what we are teaching or how we might do so in empowering ways but rather by the experiences and beliefs that shape us into who we are and the immediate meaning it has for any work we might do.

As a Haitian-American, I (Kathleen) have faced many instances of not belonging to the majority group. One incident specifically comes to mind. I was an entering freshman at a state university in a suburban town in the Northeast. During the orientation preceding my freshman year in college, I had taken an exam along with the rest of the incoming class to determine in what level of math I would be placed. Out of the three possible scores (1, 3, or 5), I scored a 5, which meant that I placed out of freshman math and would not need to take that extra semester of it. When I presented my score to my academic advisor, he questioned the validity of the score. When I assured him that I had in fact scored a 5, he advised me, "Go ahead and take the freshman calculus class anyway. It will probably be closer to your level and you don't want to do poorly in the higher level." What did I know? I trusted this man, whom I believed had my best interest in mind. He was someone who was older and therefore I assumed he was "wiser." It was clear to me in retrospect that this was not the case. I can't be sure if his recommendation was based on the fact that I was black or a female; in either case, I was sure it was biased. At the time, I did not feel confident enough to stand up and say anything.

It was very clear that I was in the margin of that orientation experience. I was a woman, a minority, and did not have the resources to know how to handle this occurrence. Reliving circumstances like the one with my advisor seemed to cause an internal struggle for me as to where and how I would fit in with the population I was teaching. This was because at times I felt as if I could relate to the struggle of people who are traditionally marginalized, yet at the same time I wondered if those groups viewed me as an insider.

I began research as a teacher-researcher in a homeless shelter in the Northeast. The shelter was a long-term housing site for families who did not have the financial means to make it on their own. Most of these families were African American. As I began teaching science to the prekindergarten class there, I was struggling to "find my place" with the kids I was teaching. Despite the initial nervousness of getting used to the children and becoming comfortable with the material I was teaching, I realized that I soon felt like I was one of them. This was probably easier than it may have otherwise been because of the fact that the majority of the children were African American. Because of that, there were certain cultural values and beliefs that I could understand not only because of the fact that I understood much of the African American culture, but also because most people considered me to be African American.

When I started to struggle with the insider-outsider theory, however, was when I began research in a large southwestern city with a high Mexican population. In this setting, I felt very different from the children whom I taught. This feeling was mostly due to racial differences. There were other differences between the population in the Southwest and the one in the Northeast; the housing site in the Southwest was for victims of domestic abuse and/or violence who were also experiencing financial hardship. The children at this site were also much older (upper elementary) and a lot more aggressive vocally as well as physically. It was here that I began to think more about who I am as a researcher and how I am viewed by the various populations with whom I work.

Working as a beginning teacher-researcher in poor urban settings has been one of the most eye-opening experiences I (Kathleen) have ever had. Not only was this one of my first experiences in science teaching, but also it was with a population of children that is usually considered at-risk and difficult to teach. I might agree that this experience was difficult. However, it was not for the reasons that one might assume, such as unsafe conditions or lack of teacher or student motivation. Instead, translating theory into practice was difficult for two reasons: First a great deal of critical theory and multicultural education writings are just that – theories. As a result, I had to learn to construct my own vision of what these meant in the classroom. Second, sometimes these theories didn't play out like they say they would and I had to theorize on my own.

Kathleen's questions about insider and outsider are central to my own thinking but for different reasons. I (Angie) am a white Italian American who grew up in a middle-class family, entered the sciences, then left bench work for education. My reason was that I felt alienated in the lab, in part because there were few women in my field and in part because I felt as if I always had to leave part of myself at the lab door when I came to the lab to do science. I moved from a general interest in equity issues and feminism in science education to a specific interest in the intersections of race and gender with poverty in science education after having spent nearly three months homeless with my husband and dog. The experience left me reeling as I struggled to balance preparing to teach with finding adequate privacy, places to clean, and places to sleep. Despite the difficulties we faced while being homeless, we knew it could have been worse. For example, I know our cultural capital as educated white people got us things we may not have gotten otherwise. We would enter a fast-food establishment and go directly to the bathrooms to wash up. We were never harassed for using the bathrooms for this purpose, even though we never purchased food from these establishments. We also went to nearby university's gyms early in the morning before they checked ID cards to use their showers. We were never once questioned about our presence at these locations—even when we had to ask officials there questions about where the locker rooms were!

The intersections of race and class continue to teach me to be more aware of what exactly cultural capital is. I have been teaching after-school science in homeless shelters for five years. To my surprise, about two-thirds of the way through one school year I learned that many of the youth at one particular shelter believed me to be "Latina." When I informed the youth that I was "white," the youth were surprised because they said that I "did not act or talk white." While statements like "talking or acting white" were surprising on one hand, on the other hand they seemed appropriate. The children were reacting to what their lived experiences had led them to believe, for instance, that there is a "white" way of talking and acting. Because I did not fit their understanding of acting and talking a certain way, they allowed me into their community to share and learn with them. It is clear, from their confusion of my ethnicity, that teachers are not the only ones who bring cultural assumptions to the classroom. The students also have notions, based on experience, that they bring to the classroom.

The question that this postcard raises is related to Delpit's (1988) idea of "cultural conflict." It is especially obvious with children in urban poverty that most of their teachers will be in conflict culturally with them. For us, as science teachers, the cultural conflict we experience with our students plays out in many ways, from how we might ask them to engage in an activity such as clean up to how we address each other in the classroom or playground. One of our biggest challenges comes in deciding whether we recognize their life-worlds based on what we think they are telling us or on what they are actually saying. In other words, with our cultural assumptions, it is often difficult to recognize the ideas, values, and beliefs of the children without first examining our own ideas. Villenas (1996) discusses at length the need to not only examine our situated positions in society as teachers and researchers, but also how this position influences how we learn to understand the world around us. She argues that if our subjectivity as teachers and researchers is not examined, we inevitably participate in the perpetuation of oppressive structures in society. In schools, the oppressive structures would include not acknowledging our own subjective positions and the kinds of understandings those positions advocate as well as the lifeworlds of students. Therefore the main question this postcard raises for us is, How can teachers ensure they are recognizing and using the ideas, beliefs, and experiences of their students in order to make science emergent and transformative?

Rethinking Postcard #1: Uncovering our own Assumptions

One way that we can teach in the interface is by addressing the issue of cultural conflict raised in the first postcard. As science educators, we must understand where we come from and stand culturally in relation to our students. The conflicts of race, gender, ethnicity, and socioeconomic conditions may be the factors that decide whether we are accepted as insiders or outsiders by the group with which we work. In an article in *Educational Researcher* (1998), Banks introduces a *typology of crosscultural researchers* from the racial-ethnic perspective. They are indigenous-insider, indigenous-outsider, external-insider, and external-outsider. This typology comes from the idea that "individuals are socialized within ethnic, racial, and cultural communities in which they internalize localized values, perspectives, ways of knowing, behaviors, beliefs, and knowledge that can differ in significant ways from those of individuals socialized within other microcultures" (Banks 1998, 7).

In urban science teaching, and especially with the homeless population, children are part of the minority, or margin. In most of the public schools they attend, they are taught science from an outsider's perspective. The fact that science is not usually related to their experiences or culture is one of many factors contributing to why students of this background often feel disconnected to science and do poorly in it or develop negative feelings towards it as a result. We have also felt the impact of being marginalized either in education or in the field. Because of that, we did not want to teach them in a way that focused on our differences, but rather in a way that embraced the experiences that we, the teachers, and the students have each encountered. However, we also come to teaching with assumptions, ideas, and beliefs based on the experiences we've had.

The indigenous-insider perhaps sums up where I (Kathleen) believed I was in Bank's typology. Because I am a black woman who has grown up in urban communities, I believed that I would be viewed by black or Hispanic homeless families as someone who understood their culture and could relate to and perhaps even speak for their community. However, this naive assumption soon began to cause some unrest. When I began teaching in the Southwest, it was especially evident to me that I was an outsider to the children involved in the after-school science program. For example, for much of the time that I was teaching, many of the children called me "April." I wasn't sure why they were calling me by another name or where it came from. I soon discovered that April was another one of the few black women employed by the shelter. The children thought I was this same April! Yet they never appeared to have any problems distinguishing the numerous Hispanic women who also worked at the shelter. This confusion reminded me of my position as an outsider.

In relating her experiences in college, bell hooks (1994) recounts professors in her feminist classes using the term "women" to make the experiences of white, privileged women a norm. By using women as the norm for only a certain group, countless other cultures, races, classes are left out of the picture, in essence made invisible. The invisibility that bell hooks discusses relates to the assumptions that we and the children brought to the shelter. I (Kathleen) was using my experience as a minority as the norm to relate to the homeless children. The children, on the other hand, were using their experiences to label me as another black woman employed at the shelter and Angie as "Latina" because she had light skin and didn't talk or act white. The idea of cultural conflict is not one-way. It is obvious that the students themselves bring experiences that shape the nature of their interactions with their teachers and their learning. It was also evident that even though I may have presumed that because I was a minority I could relate to them, it proved not to be so for them.

Postcard 2: Teaching the Canon: Purple Polar Bears

The homeless shelter where we taught accommodated a prekindergarten class. The regular teacher in this class was Ms. Marie. Ms. Marie had her own teaching style, which seemed to incorporate a great deal of memorization. Her reasons were that she believed there were some basic things these children needed to know before they entered "real public schooling." She emphasized things such as being able to say the days of the week, listing the colors, and counting numbers. She is what some would consider a traditional teacher. Quite often, for various tasks, she asked the students, "Who can remember?" while she waited for an answer. She appeared slightly frustrated if the children could not remember what it was she had repeated to them several times the day before, and her frustration would be visible in her bodily expressions and her talk. She was also highly structured in her daily planning. Each day began with the children sitting quietly in their seats. Once they were quiet and kept their hands to themselves, breakfast was served. After breakfast, the children (twenty-five of them!) were asked to sit quietly on the carpet on the far side of the room by the windows. During this time, Ms. Marie would have the children sing songs. She would also ask them questions about the things they had learned the day before, such as, What color is this? What day is today?

We were often present during this time because science always followed. We taught science in this class twice a week. During the rest of the day, the children were taught by their usual teacher, Ms. Marie, and between two and three teacher's aides. Science was not taught any other day except for when we were there (along with several interns who came from the university with us). There were some exceptions, however, and one of those exceptions is the focus of this postcard.

One of these exceptions was when Ms. Marie continued our theme with animals on her own. One of the activities she engaged her students in was coloring two large polar bears. Each bear was about three feet long and two feet tall. They were both drawn on butcher paper. The pictures were placed on two big tables along with cotton balls, glue and paint. The children were then asked to paint the polar bear. The children glued some cotton balls on and used some of the white paint provided by Ms. Marie. However, they also opened up their other paint supplies, and by the time the class period was over, the students had made their polar bears every color imaginable under the sun! The bears were hung on the wall in the hallway outside the classroom. The next day, when we arrived at the classroom, we found one of the teachers' aides out in hall painting the children's polar bears white. When we asked the aide why she was painting over the children's work, she stated that Ms. Marie asked her to do that because she did not want the children to really believe that polar bears could be purple! By the time the children entered the hallway for recess, the polar bears were completely white.

Our initial reaction to the purple polar bears was to quickly judge Ms. Marie for diminishing—marginalizing—the children's creativity and efforts. However, such a quick judgement robs us (and our future students), as teachers, of what we might really learn from this episode. Thus, we see several ways we can read this story. On the one hand, Ms. Marie wants her students to know that there are certain right answers and certain wrong answers. For example, in the citywide exams so common in the city where these children live, if the children answered a question by stating that polar bears could be purple they would surely be wrong. On the other hand, Ms. Marie presented her students with an open-ended artistic activity then rewrote their work without any questions or discussions. Besides, it is completely possible for polar bears to be purple! No context was given for the assignment, and thus context is assumed. It is conceivable that polar bears in captivity could have their hair dyed purple. Improbable? Yes. Impossible? No.

The question this postcard raises for us is, What happens (or what should happen) when ideas that students generate in the name of science differ from the canon of science? What kind of response does such conflict warrant? This conflict between children's ideas and the canon of science is yet another issue to consider in emergent science, especially when the canon has been built around a set of European middle-class values. If a main idea of emergent science is to teach science at the intersection of this conflict, how do we ensure that students learn the canon? We do not believe that simply building bridges between children's science and real science is the answer. After all, the canon is what is used to judge most students on scientific knowledge, but the canon, as a stand-alone entity, can be alienating and oppressive. It portrays science as value-free and always right. We believe, however, that value-free is not true. The very nature of the canon is political in terms of what constitutes valid science or the right answer. By addressing this question, we can hopefully elucidate the many complex layers involved in teaching truly emergent science.

Rethinking Postcard #2: Changing the Canon

The second postcard gives us another way to look at how to teach in the interface. The canon is a powerful and political entity that is often disconnected from the lives of children in urban poverty (Stanley and Brickhouse 1994; Apple 1993). What we feel challenged by is how we, as teachers, and the children co-construct science and how we connect our lives and our students' lives to the gatekeeping canon of science. One of our goals as science educators is to broaden our definition of science as well as our students' definition of science. Included in this aim is the attempt to change the balance of power concerning who defines what science or valid knowledge is. In order for the current power tables to be turned, the voices of many underrepresented people need to be heard. This power struggle between people in the margin and those in the center is evident in science education. Students' values and experiences lead them to unique ideas about science. Delpit states, "I further believe that to act as if power does not exist is to ensure that the power status quo remains the same. To imply to children or adults that it doesn't matter how you talk or how you write is to ensure their ultimate failure. I prefer to be honest with my students. Tell them that their language and cultural style is unique and wonderful but that there is a political power game that is also being played and if they want to be in on that game there are certain games that they too must play" (1988, 292). But these ideas need to be more than unique; they need to be acknowledged so that students feel empowered enough to transform the canon as a result of their ideas. We believe that in order for this to happen, students must first be secure in what they understand science to be, as a result of their experiences, whether it matches the canon or not. However, the students must also have sound knowledge in the canon of science. As Stanley and Brickhouse (1994) posit, it is only when traditionally underrepresented students are adequately educated in and about science that will they be able to assist in turning the power tables. However, this responsibility lies in the hands of not only the students, but also the teachers and the people who make decisions in the canon.

The case of the polar bears elucidates this issue of power and canon. We are not contesting the color of polar bears. Most really are white. But perhaps it is not so much what we learn from this episode about what to teach or how to teach it, but rather the kind of framework this episode gives us for reflecting on why we do what we do in classrooms. It helps us to see that one of the issues we face as teachers is the tension that exists between the perspectives of what we want the children and teachers to come away with (the interface), the perspectives of the established practicing teacher (traditional science), and the perspectives of the children themselves (life-worlds). For example, Ms. Marie could have used this opportunity to study the colors of animals and how their colors are important to their survival (i.e., chameleons, lions, and insects). She could have also used this occurrence as a way to understand where the children were coming from and what in their experiences, if anything, led them to want to paint the polar bears purple. This is the interface, the place where the children's experiences meet content knowledge. The interface always exists; it merely needs to be recognized as an important component of education.

Postcard 3: "Scientific" Experiences— Tanya's Painted Nails

We had been teaching a unit on the environment and how it is affected by and affects the students. In one lesson, the students went out to a park and collected various items (leaves, twigs, dirt, etc.) in order to create a mural. One particular student, Tanya, collected much the same items as the other students. When we went back to the room, however, she not did use the material she collected to paste upon the mural. The first thing she did was to write her name on a portion of the paper. She then had me (Kathleen) trace one of her hands near her name. She attempted to trace her own hand but couldn't and asked me to do another one. She then took the marker and began to paint her nails. I asked her what she was doing and she replied that she was painting her nails orange. My initial instinct was to tell her to not do that, but instead continue working on the natural science mural. Instead, I allowed her to continue engaging in her activity. After doing her nails, she painted Angie's nails as well. She told me that after she had done Angie's nails, she would do mine. She followed through and what ensued was interesting. Other students, who perhaps started to notice the attention she was getting by doing this, began painting their nails or those of the students around them.

Our initial response was that we needed to validate Tanya in what she was doing; she probably was imitating something she had seen her mother or her mother's friends do. Yet as science teachers, who allowed this activity to transpire in our class, we forced ourselves to think harder about this event. To the outside observer, Tanya's activities might not be perceived as scientific. Even our backgrounds in science and science education made us question how what Tanya was doing constituted science. However, we have to remind ourselves that as science teachers our role should not be to relate every activity in which students may engage back to science. This would, in effect, be doing children a disservice because it would suggest that their activities are only validated by the science label. The idea here again is not to validate the activities with a scientific label, but to validate the activity alone. In fact it is clear that there is opportunity in the classroom for many activities to not be scientific in nature. We want the students to feel validated with the ideas they express; however, everything they do is not science. Yet we are stuck by our own ideas about science and the ways in which our ideas are bounded. In other words, our own beliefs, ideas, and experiences frame our definition of science and what and how we label activities as scientific. This postcard raises questions about validity and the transformative aspect of emergent science: How does the nature of our response generate the kinds of spaces needed for students to challenge their own and other children's ideas. What are the ways in which we use students' experiences to construct new science?

Rethinking Postcard #3: Responsive Teachers, Transformative Students

The question of validation is a difficult and complex one. On the one hand, as science teachers we struggle with not getting involved with all the other "stuff" (i.e., all the possible tangents that students may attempt to follow). On the other hand, we realize that that stuff is crucial to children's development as confident and empowered students as well as to their understanding of science and our understanding of them. Tanya's painted nails give us a lesson on why teaching in the interface is so important. Her actions speak to us about the importance of responsive teaching. The validation of Tanya's actions was crucial, but perhaps not enough. Perhaps we should have responded to the actions of Tanya by redefining how we initially envisioned the science activity to reflect her experiences. In science circles, science is described as ways of knowing and coming to an understanding about ourselves and our environment through careful study, such as observing, thinking, experimenting, and validating; a process for producing knowledge; a blend of logic and imagination; and a social activity (AAAS 1989). If we talk about science as emerging from students' life-worlds, then we have to be open to what their life-worlds have to say to each of these components of science even when this challenges our own ideas.

For example, we didn't want to label Tanya's fingernail painting as science, but on the other hand, we think there is power in embracing her response to our prescribed activity. In painting her nails and influencing the other children to follow suit, there is no question that Tanya used her logic and imagination to turn her actions into a social activity. She even observed, thought, and experimented in her actions. Perhaps instead of having the students write just their names on the mural, we could have followed Tanya's lead and integrated the exploration of the natural world around them (i.e., the twigs, leaves, and grass) with their social world (i.e., painting their nails). This may have led to further discussion on hands or nails and the types of materials we use to adorn them and how those differ (i.e., magic marker versus nail polish, components of various types of jewelry). Furthermore, we could have begun to talk about the differences between their hands, which might have led to a study on anatomy and the differences between the "hands" of different types of animals. If we question ourselves as teachers and our responses to the diverse children we teach, if we question the values embedded in science, and if students are given the space to look at their ideas in new ways, something dynamic can happen in the classroom. Students can transform their teachers, their peers, their own ideas, and science.

Discussion

The postcards we have presented aim to expose the questions we face in our science teaching: What does it mean to do emergent science with children in urban poverty? How is emergent science transformative, and for whom is it (or ought it to be) transformative? The issues brought up in each of the postcards all connected to the idea of emergent science and transformation. In the first postcard the question that was raised was one of cultural conflict. How can we uncover our own cultural assumptions enough to understand students' values and experiences? The second postcard brought up the issue of the canon. What happens when the experiences that children in urban poverty share from their life-worlds conflict with the traditional science? Can our work in classrooms transform the canon of science? Finally, in the third postcard we raise the question, How does the nature of our response generate the kinds of spaces needed for students to challenge their own and other children's ideas? What are the ways in which we use students' experiences to construct new science?

These questions brought to mind some reflections recorded by one researcher on our project. We relate this reflection because it portrays our attempts to construct meaningful scientific experiences that are emergent and transformative. That reflection follows.

During the after-school science program, we had been involved in studying all kinds of insects. One particular outgrowth that the kids came up with was to make a home for future caterpillars. We decided upon a butterfly garden. On the day that we were to build the planters for the garden, I (Tanahia) was supposed to be assisting Claudia with her planter (which eventually turned into a desk making project-more on that later), but we had no hammer so we sat outside and waited until one was available for us to use. During that time I observed Junior hammering away and somewhat resisting any help from the new Child Activities Coordinator. She was there on the ground attempting to hold the pieces of wood together for him as he put the screws in and hammered them. A few times the nails bent or went through the side of the wood but Junior continued to hammer the pieces together, sometimes using more nails than necessary (by my estimate). There was an eagerness to work on this project, to see it through to completion. At times I noticed Junior being quite the perfectionist. When he turned the planter upside-down to put the bottom on it, he exclaimed, "Aw, damn, it doesn't fit." He called out to Angie for help moments later after he had tried tirelessly to rearrange the two pieces of wood so that they would fit across the bottom of the planter. Angie assured him that it was OK to have a small opening because the water would have to drain out anyway. This seemed to calm Junior as I witnessed him continue his project with as much zeal as he had begun with. After I had gone into the portable with Claudia to work on her planter I returned outside to find that Junior's planter had fallen apart. I was almost in panic mode although I didn't show it. I wondered what Junior would say about his project seeing it in pieces on the ground. My tension was somewhat eased when he returned to the grassy area and asked Jason (very calmly) what he was doing. Jason explained that his planter had fallen apart. I don't know if Junior was angry, disappointed, or indifferent about what had happened, because at this point there was a disturbance when one of Rueben's friends from school, accompanied by his father came on the premises, which is a absolute violation of the shelter's rules. Evidently Rueben had invited the boy to play on the computers and his dad brought him over after school. When the children heard that "the boy from school" was there many of them (the older ones: Maria, Claudia, and Junior) left the area where we had been working and ran over to greet their schoolmate. I had gone outside to look for Claudia, who had bolted out of the portable minutes earlier after Maria had come to tell her about the boy, when I observed something that sort of made me angry and sad all at once. One of the child service workers had come over to reprimand the children. I understand that she was taken aback by a man and his child showing up at the shelter to "play on the computers" but the way in which she handled it was inappropriate in my mind. She came over in the middle of the activity and told Junior (he had returned by this point) that they are not supposed to have "just some random man dropping their friends off (at the shelter) to play on computers." I wondered and still am wondering why she was so angry with Junior since it was Rueben who had invited the boy. Although she had come to tell Faye what had just taken place, she began reprimanding all of the children and telling them over and over "you are not to have your friends over after school" "that is not allowed". I was hurt because I then associated the desk situation with Claudia with what I had just observed and wondered what type of effect the shelter will ultimately have on these children.

At the beginning of the project Claudia had obtained permission from Angie to use her wood to make a desk. Claudia then summonsed me to the portable and told me which board to hold while she hammered. She wanted this desk to be perfect. At one point she told me to hold a few pieces together so that she could get a chair and sit at it to make sure it was the right height. At times when she had trouble driving a nail through she would say, "Here you do it so it will stay together." When we finished we still had three pieces of wood left. Claudia jumped up and said, "Now we can make a bigger one!" We had trouble however, trying to fit the second half of the desk together. Consequently she had two small desks instead of one large desk. I illustrate this point to show a relationship between some of the things that the children have both inexplicably and explicitly said and done. For instance, Junior once suggested that as a project we make a playhouse. A few weeks later Maria returned to the idea of the house. Although I hesitate to analyze, I would like to ponder their verbalization. What the children at the shelter are saying is that they want a normal life (however they would describe normality). Claudia tells how she wants a desk to sit at while doing her homework and is even willing to hammer boards together herself in order to have one. Junior and Maria want to build a house that they can play in. And Rueben wants to invite a friend over after school. These are all very normal childhood desires and also very attainable if the children were not at the shelter. But they are. And even simple pleasures have to fall by the wayside in attempts to be safe and sheltered. I have never fully thought about the drawbacks of sheltered housing until I began interacting with these children. Surely it is much more important to be safe from harm and sheltered than it is to have a playhouse or a desk, but what happens when these children are older and have never had the opportunity to develop autonomy: when they are so used to being acted upon and have never had the opportunity to form close bonds with others because of either having to keep their residency confidential or for having to move away from the friendships they have formed while at the shelters. For some strange reason I think that the afterschool science activities give the children something. I don't know what it is. Maybe a caterpillar, a bug or a worm to take care of in the absence of a pet; maybe a planter of their own in the absence of a garden; maybe a playhouse in the absence of a backyard . . . Whatever it is I have begun to see this situation in a new light. Now that I reflect on it, I think this is why Junior took so much pride in the construction of his planter. It was something for him, by him when little else is.

Implications

For us, as we reflect on these postcards and reflections, what becomes clearer is important to understand how to make "science for all" emergent and transformative and a reality for children in urban poverty. We believe that one way to do this is by understanding the experiences children in poverty have had, understanding our own experiences, understanding the worldviews embedded in the domain of science, and, finally, by making sense of the conflicts which emerge when our collective experiences are brought to bear on doing science. We call this connection – this space of conflict—the *interface*, because it is where science and children's and our own experiences overlap and where children's science emerges and becomes transformative for the children, for us, and for the science we experience together. In what follows, we examine the ideal of teaching in the interface to make sense of what emergent science might mean as well as its transformative nature. We draw on the questions raised in our postcards to help us do so.

There is no question that the lives of children in urban poverty are complex and difficult. We realized that the issues of finances, hunger, and familial responsibility are but a few parts of the life-worlds of the children with whom we worked. They may have a different understanding of their role as users and producers of science than children who are not in urban poverty, but this difference is because their experiences are diverse and their life-worlds are not usually included in traditional science education. Science is at times believed to be value-free and universalist in that traditional scientists do not make issue with or attempt to address the cultures of students. However, we argue that value-free science is not possible. Science is embedded with the values, beliefs, experiences, and cultural assumptions of those who create it. Teaching science in a way that is emergent derives its strength from acknowledging the diversity of ideas, values, and experiences of children. By teaching emergent science, the lives of students can become an integral part of science education. Rather than teaching a removed body of knowledge to students, we wanted to use the experiences that students bring to the classroom to co-construct scientific knowledge. This more than adopting the ideals of the traditional constructivist approach, which doesn't question science and the types of experiences and values that children have. Instead, by bringing students' experiences to the classroom, if done and acknowledged correctly, necessarily transforms science. We admit that this transformation would be hard to describe or recognize because we don't know all the experiences of students. The difficulty in recognizing the transformation is especially true for children in urban poverty who are largely and traditionally left out of the body of science.

Teaching science that is emergent can also be transformative for students, teachers, and the domain of science as well. Children of urban poverty can learn to look at science differently because of emergent science. It allows them to be empowered because their life-worlds are used in the teaching of science. As a result, students may feel encouraged to be more involved in science in society. Moreover, we believe that as we teach emergent science, we will be transformed as well because of the relationship between science education and the lives of the children with whom we work. In order to do this we must first realize who we are as teachers and how it affects how we teach. This reflection serves to give teachers the opportunity to observe whether they are truly recognizing the values and beliefs of their students. It is this reflexive process that can lead to the transformation of emergent science. As teachers, if we are constantly reflective of our teaching patterns, we can continually work on ensuring that the science we are teaching is emergent. The example of the purple polar bears illustrates the importance of this reflexive process that is necessary for science to be both emergent and transformative. Ms. Marie prescribed an activity with no constraints. In the end, however, what the students produced was not what she expected. This would have been an excellent opportunity for her to initiate a discussion with the children about what they had created. She could have used this opportunity to listen to the ideas the students had about the way they painted the polar bear, as a method of teaching. Utilizing opportunities such as these is the key to teaching emergent science. Not only would the students feel validated to express their ideas in the future, but Ms. Marie would also gain insight into the experiences and beliefs of the children in her classroom.

Finally as aforementioned, emergent science is transformative not only for students and teachers, but for the domain of science as well. The canon of science can also be transformed as students, working with their teachers, realize their roles and utilize their power to make changes happen. Thinking about the various aspects of emergent and transformative science will get us closer to the goal of teaching in the interface. Science has traditionally been perceived as an exclusive domain. In a society that operates on a margin-center continuum, those involved with hard science have maintained a great deal of power and are therefore positioned at the center. This leaves children of urban poverty in the margin of science. However, the power that this body of knowledge has in the classroom can be diminished the more we transform the teaching of science to include the ideas and beliefs of the children we teach and who also teach us.

The children who live in urban poverty and science are integrated in many ways. However, they may be integrated in some ways that may not be apparent to traditional teachers. The fact that teachers from other cultures may not recognize children's life-worlds has never been clearer to us than with our work with homeless children. As science educators we continually seek for ways to find the connection with the experiences of traditionally marginalized groups such as children in urban poverty. That is, we seek ways not only to recognize and understand the life-worlds of the children we teach, but also to acknowledge these experiences as valid so that students transform and feel empowered enough to allow their ideas of science to emerge. We realize that if we can teach science in a way that is emergent and transformative for students and science educators alike, we may be able to do more than bridge student science with real science; we can teach in the interface.

References

Apple, M. 1993. The politics of official knowledge: Does a national curriculum make sense? *Teachers College Record* 95 (2): 222–41.

American Association for the Advancement of Science (AAAS). 1989. *Science for all Americans.* Washington, DC: American Association for the Advancement of Science Press.

Banks, J. A. 1998. The lives and values of researchers: Implications for educating citizens in a multicultural society. *Educational Researcher* 27 (4): 4–17.

Barton, A. C. 1998. *Feminist science education.* New York: Teachers College Press.

Barton, A. C. 1998. Teaching science with homeless children: Pedagogy, representation and identity. *Journal of Research in Science Teaching* 35 (4): 379–94.

Delpit, L. D. 1988. The silenced dialogue: Power and pedagogy in educating other people's children. *Harvard Educational Review* 58 (3): 280–98.

hooks, b. 1994. *Teaching to transgress: Education as the practice of freedom.* London: Routledge.

Rodriguez, A. J. 1997. The dangerous discourse of invisibility: A critique of the national research council's national science standards. *Journal of Research in Science Teaching* 34 (1): 19–37.

Stanley, W. B., & N. W. Brickhouse. 1994. Multiculturalism, universalism, and science education. *Science Education* 78 (4): 387–98.

Villenas, S. 1996. The colonizer/colonized Chicana ethnographer: Identity, marginalization, and co-optation in the field. *Harvard Educational Review* 66(4): 711–726.

Chapter 5

A Paralogical Affirmation of Emotion's Discourse in Science Teaching

Michalinos Zembylas

This article is a set of experiments in paralogy. Stories of forces that unify paradox and emotion. I want to hold infinity in my [careless] mind, and conspire with bees, and shells and clumsy cells to "touch the sky" (Jimi Hendrix) and gather all the deceiving selves to dream an [alien] grain of sand.

Michalinos: What do you mean when you say that you "use emotions in your science teaching?"

Catherine: [I describe] how I felt about something we saw or read about, how the room is looking, or how I was trying to figure something out and was having trouble or got frustrated about it. I would tell children, "I don't understand; *this* says *this* and *this* says *that*. Ha!" That's a feeling, a puzzlement. I talk to them about feelings about my family, about things that happened to me outside of school, about something I learned and got me all excited. I encourage them to share their feelings, too, if they want to, and usually they do. I encourage them to experience and feel science, and of course everything else we study.

Michalinos: But what happens when one doesn't like science or one is even scared of science?

Catherine: What I want to suggest is that we need to make sure we find something exciting and fun about something that seems boring or unreasonable to do. So even if there are not initially some exciting aspects of something, basically it's up to you and your students to find ways to make it exciting. Even if you don't like it or you hate something. How we feel influences the things we study in our classroom. If the kids really don't feel about exploring something and if my excitement and my enthusiasm at some point doesn't pull them along then we end up usually dropping that topic. We take turns pulling each other along. I don't see how you can separate emotions from learning or teaching. I don't think it's possible. Is it?

As we enter the new millennium, we realize how much we have excelled at rising technical rationality to levels that were unimaginable. Our educational systems have been built on a set of logical rules and standards. Our are curricula built on analytic and reductionist reasoning, a formula-driven mind-set, and rote memorization of endless facts. We have used these rules to perfect our students' performance in tests and prepare them to become by-the-book professionals. We want them to be highly successful in achieving high scores in tests, not to be creative, sensitive, and caring. And we have discovered that this is not enough. Our teachers feel overworked and undervalued, our students are bored and uninterested, their creative spirit is ignored. Our model of rationality may have produced a technologically advanced society, but our senses, hearts, and intuitions, in essence, our humanity and sensibility, have lost their value in a competition won by the neck up. Somehow, models such as this have missed a lot of what matters most in teaching and learning.

In everyday life, even when we teach, we do not only do one thing but we play various games and express a polysemy of "forms of life" (Wittgenstein, 1953). One of these forms is our *emotions*, a form of human activity which cannot be treated in isolation from other forms of human activity. *This is an essay about the value of elementary school teachers' emotions in their teaching of science.* It is about reflecting on the importance of teachers' excitement, joy, caring, wonderment, awe, anger, fear, guilt, sadness, embarrassment, hope, and many other emotions when they teach science to young children. What is the role of teachers' emotions in shaping their perceptions, decisions, and values, in making choices on what and how to teach science, in determining what seems important to explore with their students? This is the question at the heart of this essay. Since virtually no work has been done to explore elementary school teachers' emotions in their science teaching and pedagogy, I will attempt to show the need for such research. Therefore, this work raises a question that goes beyond the meaning and interpretation of teachers' emotions: How can teachers' emotions become a legitimate topic in science education as well as in our efforts for science curricular reform? Further, how can educational institutions (universities and schools, for example) and elementary school science teachers themselves support their personal and professional emotional growth?

The Genesis of a Study

> "I want to understand," you exclaim. Your cry looks into your face. To understand is to make yourself understood. You're afraid of your face. You take pictures, name the facts, yet are paralyzed by them. You're a witness of passing facts. Your dreams rise through the smoke of burning memories. Now, do you understand?

This essay is inspired by a three-year ethnographic study that I conducted with Catherine Myers, a 25-year veteran elementary school teacher. In a sense, this study reflects my own journey through emotional landscapes frequently invisible, but it is also an exploration of the landscapes of elementary school science teaching and its emotional ups and downs through the eyes of Catherine.

My motivation to study emotions in elementary school science teaching began when I realized at some point in my own teaching how my thoughts, my actions, and decisions were suffused with emotion. I became fascinated with how the excitement and passion I felt at some science topics (astronomy and space science, in particular) transformed my teaching and gave me unimagined joy and hope in ways that I thought I was able to offer valuable and fascinating experiences to my students. Other times, I puzzled when my boredom with teaching some mandated curriculum topics infected my own students and I felt that I wasted their time. Still other times, my lack of science knowledge made me feel incompetent, worthless, frustrated, and angry at myself because I felt that I was unable to do my job adequately. I felt that science was so complex and my knowledge of it was always flawed and partial. I was dreaming for the day that I would walk into my classroom without feeling uncomfortable but competent and knowledgeable.

Now I know that day will never come, partly because my knowledge, whatever form that may take, will never be complete; it will always be traveling to unknown territories, tracing dreamlike trajectories. On the other hand, I have had too many lessons on the value of humility and self-doubt and it is about time that I do something with them. These experiences and many others reveal how powerful emotions can be in shaping the course of teaching science to young children. I am still intrigued, just as I was when I first started teaching science to fifth graders, at how teachers' emotional responses, complex and multifaceted as they are, shape our everyday lived experiences when we teach. I now find the lack of attention to how elementary school science teaching is associated with emotions to be disturbing and, therefore, limiting. This is, perhaps, one of my main motivations—why I study emotions in science teaching. I want to make myself, and perhaps others, too, more self-reflective about the connectedness of our emotions and the ways we teach. For I will argue that in expressing our emotions, we reveal our personal educational vision. There is certainly truth in the saying that "we teach who we are." Jim Garrison puts it more powerfully: "We become [as teachers] what we love" (1997, xiii). In my view, teachers' emotional responses to children partly underlie teachers' existing moral judgments. For better or worse, teaching emerges from one's inwardness (Palmer 1998). By this I don't mean to imply that power relations, desires, and reasons have no place. In fact, I will argue that part of the *wholistic approach* of knowing and teaching that I advocate is that somehow power issues, desires, emotions, imaginations, and so on have an important role to play in improving the quality of science teaching.

I decided to work closely with Catherine Myers when I discovered that we had similar concerns and questions about the role of excitement, passion, and enthusiasm as well as the influence of frustration, lack of self-confidence, and anger in science teaching. I was fascinated with Catherine's interdisciplinary approach in science, and from the very first day I entered her classroom I saw an enthusiasm for science (both from her part and her students' part, too) that I rarely saw before. Catherine's focus throughout her career has been multiage classrooms, in which she has collaborated with her students to construct and implement their own in-depth, integrated investigations. Catherine has developed an in-depth integrated inquiry approach to meet the diverse needs and interests of her students. Through this thematic, integrated curricular approach, students construct their own knowledge while experiencing the connectedness between subjects, using authentic materials and activities that are compelling and interesting. Such an approach engages imaginations and enthusiasms and evokes the emotional and aesthetic responses of the children.

I did an analysis of Catherine's narratives, classroom observations, emotion diaries, and instructional materials for three years; the focus of my study was to describe how Catherine *uses* her emotion in her science teaching. Therefore I looked at the ways she expresses her emotions, what kinds of emotions she usually expresses and how she talks about them, and how she uses her emotions to plan, create, and motivate action in her classroom. One of the major issues I explored is the relationship between emotion and knowing. In fact, one may argue that emotion is one of Catherine's ways of knowing, because emotion becomes a source of evaluation and information on how she feels about various events, topics, people, and objects. For example, earlier in her career her emotions of frustration and anxiety as well as her "shame" (to use Bartky's [1990] term) about feeling inadequate as a science teacher have to do with being asked to teach an unproblematically masculinized science. (Later in the chapter I talk more about this.) Catherine frequently centers her efforts to find or construct support for her own beliefs about including emotions in her science teaching and her need to explain her beliefs to multiple audiences, including her students, colleagues, and principal.

My aim here is to show how Catherine's emotions are significant dimensions of her teaching, playing explicitly or implicitly important roles in her personal and professional development. To paraphrase Denzin (1984), one might ask, How is emotion in science teaching, as a form of consciousness, lived experience, articulated, and felt? I am interested, for example, in how Catherine combines her passion of science with critical analysis "to define and identify how and when particular emotions inform and define knowledge" (Boler 1997, 231) and how the discourses she and her students construct in the classroom determine what can and cannot be felt and/or expressed.

In this chapter, I will outline the importance of emotions in science education, and I will describe my approach in researching emotions in science teaching, providing two vignettes about how Catherine's emotions and her reality construct one another and how her emotions both inform or limit possibilities in her science teaching. I believe that such a line of research may show the multiple ways that teachers' emotions create contexts for the teaching of science. In order to identify, imagine, and construct more exciting experiences in science learning and teaching, we need accounts that recognize the pedagogical importance of how emotions shape classroom interactions. Indeed, what is the pedagogical importance of considering emotions in science education? Why should one pay attention to how teachers are emotionally engaged in the teaching of science?

The Importance of Emotions in Science Education

> You solemnly mimic the sound of paper. And then a weak letter jumps out, fixed ignorant against the stormy sky. What a madness! You are convinced you have been entirely right to stay at home.

When teachers are engaged in their teaching practices they often express many emotions, such as frustration, despair, joy, surprise, or anger, because teaching involves interactions among people. These emotions can be positive or negative and manifest themselves in different ways, fulfilling a variety of functions, sometimes perhaps through action and/or speech. The significance of emotions in teachers' work may be realized further in those events where emotions reveal a truth about a situation, a truth which rational thought processes may ignore or hide.

Nias (1996) observes that teachers invest their selves in their work and so they closely merge their sense of personal and professional identity. They invest in the values which they believe their teaching represents. The construction of self and teacher identity then is a process that involves teachers' values, emotions, and actions. Consequently, Nias (1996) points out, their teaching and their classroom become a main source for their self-esteem and fulfillment as well as their vulnerability. Research on these emotional reactions of teachers to their work should be helpful in planning, social relations, sharing, the pursuit of their personal growth and development, and the growth of their students. Studying teachers' emotions has the potential for linking teachers' personal experiences in the teaching of science with broader institutional forms of school organization. For example, the school's organization enables and shapes teachers' emotional experiences. At the same time, emotional experiences work dialectically to maintain, confirm, or change this very structure (Franks and Gecas 1992). By seeking ways to establish contexts that promote constructive, positive emotions and caring, students, teachers, and researchers can facilitate constructive action, stimulate skills and positive habits (van Dam and Steutel 1996), and bring changes to all aspects of education. I claim that *transforming* emotions (for both teachers and students) through careful exploration and refinement is both an educational and an aesthetic process. But transforming these emotions implies that first we *understand* them.

Researching the role of emotions in the ways elementary school teachers teach science may open the way to a richer understanding of elementary school science teaching, one which highlights the tacit dimensions of teachers' knowing and is sensitive to their emotions, their intellectual temperaments, their desires, and their imaginations. The implications of such a perspective for elementary school science teaching and science curriculum reform are enormous but by no means clear. I argue, though, in a Deweyan sense, that emotion matters just as much as cognition in the unity of our experiences. A major idea running through this chapter is the reclamation of the interrelation between emotions and cognitive phenomena, and, in fact, the inability to demarcate their boundaries, both philosophically and practically. We may appreciate the power and significance of emotions when we see them as an *experience* of knowing, one that emerges to illuminate the affective character of our human existence, and that points to the power of the unity of our experiences.

By *understanding* in this context, I also refer to Gadamer's (1975) notion of a "premordial mode of being" of human life itself. Understanding is no longer a cognitive activity but it has ontological dimensions. In other words, understanding is a mode of experience. Dilthey (1958) and Weber (1949) called this *Verstehen*, the empathetic understanding or imaginative reliving. Similarly, emotions are not only ways of knowing but ways of being in the world, and teachers' emotions are ontological ways of *being in the classroom*. This ontological perspective opens up new possibilities in relation to which educators may come to terms with the teaching of science. It is precisely in and through an understanding of teachers' emotions in elementary school science teaching that we can come to a more sensitive and critical understanding of emotion, teaching, curriculum, and pedagogy. As I see it, I think that a different ontology is needed in science education, one that views emotions as not distinct from a disenchanted reason and views teachers as wholistic beings. Of course, such a non-dualistic ontology implies a different epistemology. In other words, we can know the world with both our emotions and mind because we are of both emotions and mind, because we are shaped by them, and because we shape them.

The Apparition of Modernity in Science Education

Plural, friable, sensual. Without casual recollections of seasonal infinities. Shifts, leaks, flows. With inexhaustible visions of eternal temporalities. The voyages through timeless seasons transfigure my lives.

My research on emotions is an attempt that seeks to come to grips with what I take to be a fundamental crisis in science education of our time: the fact that many science educators and researchers still pay allegiance to *modernity*, namely, a preoccupation with the rationalist roots of science teaching, one that emphasizes the existence of clear beginnings, sequential orders, and predetermined endings. Following Stephen Toulmin in proclaiming that "an exclusive preoccupation with logical systematicity has been destructive of both historical understanding and rational criticism" (1972, vii), my own deeply held conviction is that the preoccupation with researching teachers' thinking and behavior has been detrimental to understanding teachers as *wholistic* individuals with emotions, histories, and rationalities. Understanding teachers' emotions can be a step towards a richer understanding of teachers themselves.

To put this in a wider historical context, I think that the history of science education in the United States is a remarkable story of conflicts of purposes, marginalization of teachers, and prevalence of rationalism in deciding who, what, and why certain movements and ideas should prevail in the education of young children in science. Surprisingly, most of the current conflicts are derivative articulations of past positions that reappear under a new mask. For example, DeBoer (1991) identifies three major themes that have often recurred in science education: teaching the science disciplines as structured bodies of knowledge to be learned as logically organized subject matter, teaching science as a set of investigative processes, and teaching science as a human activity closely interconnected with its technological applications and with the rest of society. However, DeBoer clearly shows how the teaching of science as a structured body of knowledge has been the most dominant focus of school science curriculum during the past one hundred years. Science is taught as a rationalist, intellectual endeavor of finding the most appropriate concepts and theories to describe and explain our world. And part of good science teaching has required teachers to think about their work in terms of the great dichotomies: either science facts or feelings, rigorous and objective scientific evidence or personal and biased emotions (demons, as many philosophers describe them). Ironically, more often than not teachers were left out of these debates and the decision-making process of what and how science should be taught.

School policies and structures are often skillfully organized to prevent teachers from inventing their own thoughts, from feeling their own emotions: "carefully" written curricula, science-in-a-box, must-teach science topics, proper professional behavior (e.g., remember who has the power in the classroom), and so on. Obedience to the experts (whoever these are) becomes the only true unimaginable and apathetic act, the only act of lack of emotional, ethical, and political knowing. Evidently, it seems that some curriculum developers, most textbook publishers, and many school districts have formed an "unalterable conclusion upon the nature of what is to be known" (Woolf 1931, 218). I find these educational arrangements that aim to crystallize teachers' (emotional or not) own inventions too cold, too universal.

Therefore, to study the role of emotions in teachers' lives in relation to the power relations in the classroom and the school culture more generally is fascinating for a number of reasons. First of all, science education has not been explored in terms of its relationship to emotions, either in theory (pedagogy) or practice, however one wants to define these terms. Furthermore, the ways in which science education perpetuates certain moralities of compliance, self-blame, and professional behavior (for teachers) seems an excellent site for exploring resistance from teachers who survive and flourish without subscribing to such moralities. One of the fascinating things about emotion is that it permeates our language, our decisions, and actions in the classroom, our anger and disappointment with the system and so on, and thus a meta-emotion discourse reveals more complexities about norms, ethics, and morals learned within the educational context from the very moment a teacher enters it. I share Foucault's (1983) observation that the part of ourselves most relevant to morality is our emotions. According to Foucault, emotions can play this role because they are constituted as the core of one's self.

Evidently, then, my claim that curriculum philosophies in science education appeal to teachers' emotions by manipulating them and establishing certain moralities makes the study of emotions even more compelling. For example, considering that the majority of elementary science teachers are female (although most policy makers and administrators are male) there is a certain "emotional hegemony" (Jaggar 1989) being constructed by the dominant (patriarchical) values. Women's "increased emotionality," for instance, is seen as a disruption to responsible and professional teaching; such a view, I think, suppresses teachers' emotional vitality and their emotional ways of knowing. Further, women's silence is often interpreted as willing agreement to their subordination. Unless we develop the ability to envision alternatives and transformational possibilities (Boler 1999) the dismissal of emotions (i.e., their blocking) is established as a standard of morality—the appropriate thing to do—and consequently it is perpetuated.

Science education is also a crucial site to study emotions and their connections to ethics, epistemology, and ontology of scientific knowledge because traditionally (for example, with all the hype about the neutrality of the scientific method) science is rhetorically identified with the power of reason. Thus to understand the limitations and deceptions of the slogan that scientists (and science teachers for that reason) are neutral (or should be so) shows how unaware they are of the emotional components of their lives and thus makes science education a promising site of study. For I think that besides the rational truths that exist in science and in science education—as many scientists and educators would undoubtedly claim—there are also many emotional truths, hidden, ignored, and abused. The story of the famous biologist Barbara McClintock illustrates the value of feelings in doing science and the power of emotional truths. McClintock, according to Keller (1983), wants to show that science has a deep emotional aspect to embrace the organism one wants to know by feeling for it. McClintock "gained valuable knowledge by empathizing with her corn plants, submerging herself in their world, and dissolving the boundary between object and observer" (Rosser 1992, 46; quoted in Palmer 1998). McClintock's life and work indicate that intuition and emotions cannot be isolated from reason or from science.

Lorraine Code (1991) argues that examples like McClintock's life and work remap the epistemic terrain and speak of an alternative epistemology of "knowing other people" and of respecting other ways of knowing, such as emotion and intuition. Code's model acknowledges the subjective aspects of knowledge construction, as shown in McClintock's work. She points out the multiple relationships between observer and observed—the detached, dispassionate spectator is only one of many possibilities. This also implies the recognition of the vast complexities in nature and in people's understandings of it. Embracing emotion, intuition, empathy, and reason in scientific investigations opens news possibilities for enriching our understandings about nature and about ourselves. If Hochschild (1983) is right in thinking that our emotions reflect a learning relationship between us and the environment, then emotions can serve as a source of insight and creativity. There *is* room for emotions in science teaching and learning, if we realize that they are not our *enemies*, but in fact our *allies* in understanding our lives and succeeding in our world. In this sense, emotions may be *epistemologically advantageous* (Jaggar 1989) in approaching the world of science as a relation of affection, empathy, and love. Therefore, how can one ignore the role of emotions in the teaching and learning of science in our schools? It is to this end of finding out these wonderful emotional truths in science teaching that I invoke the study of emotions and challenge the current mood of controlling teachers' emotions in education.

In the rhetoric of didactic teaching about must-known scientific facts and the conservative establishment of standards for all (in the forms of rules and norms applied to all communities), one of the underlying assumptions appears to be this: science, and its faithful followers, are committed to objective, rational, neutral, and universal truths. The ultimate goal is to reach an establishment of rules that apply to everyone so that accountability is equally ensured. Such commitments rarely explore the role of emotions and would rather live with the illusion of objectivity than the acknowledgment of the emotional and other subjective aspects of our lives. The rhetoric of conservatism, of course, is important for those who want to maintain a certain control of educational affairs, because many of the proposed solutions are located in the larger society and imply a historical tradition that pays allegiance to rationalist-managerial approaches in solving educational problems. This approach has been a primary force in the field of education and in particular in curriculum development (see Popkewitz 1998).

Catherine's story about the ways that she uses her emotions in her science teaching in many ways contradicts such traditional approaches in science curriculum development. Her emotional ways of knowing permeate her entire professional life in powerful ways and the learning and teaching outcomes are tremendously affected because of that. Teachers' emotions are not an isolated aspect of their pedagogies but rather are a crucial component that brings together other dimensions of their personalities. But before I share two vignettes, from Catherine's science teaching, that show the importance of her emotions, it is valuable to highlight the theoretical framework that underlies my account on emotions.

A Nomadic Approach in Theorizing about Emotions

My eyes dazzle from the black snow. I dream of flying. My mind travels in distant skies. I think, I forget time, I ignore death. Suddenly I become a fish in a wide river with irregular rocks. What I know is always near, remembering my travels around the earth, the sea, the air.

My effort to construct an account about the different ways that one can theorize emotions in everyday life in general, and in science teaching in particular, can be described as *nomadic*. In a sense, this is similar to what Deleuze and Guattari (1987) describe as "nomadic thought," an effort that is informed by moving through various "disciplines" and rejecting artificial bifurcations such as theory/practice, emotion/reason, subjective/objective, and public/private. This nomadism has not only epistemological but also existential components, as Braidotti (1994) argues. It traces a style of thinking that is always in movement, in a stream of *translations*, to use Latour's (1988) term, of adaptations of thoughts, conditions, and emotions—including the reflection and theorization on them.

My goal is to construct a nomadic account about emotions and teaching, to invent new images and new modes of thought, and finally to evoke, perhaps, new emotions.

My approach begins by deconstructing the historical dichotomies between emotion and reason where reason is seen as masculine, analytic, cold, objective, and universal while emotion is perceived as feminine, synthetic, warm, subjective, and private. These unfortunate dichotomies may be dated back to Plato's teacher Socrates, and the emphasis on clear concepts as essential to cognition, and Aristotle's position that women's souls were less rational and more inclined to passion (*pathos*). These dichotomies were reinforced by Cartesian rationalism and the idea that rational thought is the principal tool for understanding the eternal and objective truth. Most times passions continued to be taken as opposites to rationality and, therefore, sentimentality and emotions were left to women. Kant verified these polarities between reason and emotion and separated cognitive faculties into conceptual and intellectual on the one hand and perceptual and sensible on the other (see Johnson 1990). Reason, Kant claimed, ruled over passions. For all the changes since Plato and Aristotle, Western philosophy has held fast to the polarities established between reason and emotion. Even to ask how emotion and reason interact in a specific situation is partly caught up with the old dichotomies. In my view, to explain human complexities by the division into reason and emotion is hopeless, because we consist of many selves, not only two. Our life oscillates not merely between these two poles but among many others, sometimes unknown to us. To reduce the rich complexities of our lives to a formula so simple, so rudimentary, is to see the world through the lenses of delusive formulas and artless simplifications (Hesse 1961). Once we are outside the framework that divides reason and emotion, it is possible to see the *wholistic* aspects of various components in the emotion-reason relationships, which are violently falsified by undue dichotomizing (Jaggar 1989).

Obviously then, I reject the dichotomies that have been so entrenched in modernist assumptions—between cognition and emotion, reason and feelings. Reason and emotion, I believe, are not demarcated in ahistorical ways. For instance, teachers who share the "same reasons" may nevertheless make different decisions under similar situations. I think that emotions are, at least, one of the many factors deeply involved. Therefore, the complexities of different emotions, though all constitutive of (relatively speaking) the same reasons dictate different choices by different individuals. In other words, although teachers may share the same reasons, they do not apply them in the same way. Such choosing, I claim, is not simply a rational activity but also an emotional one. Like Polanyi (1958), I argue that the tacit knowledge of science teaching (including emotions as ways of knowing) may be more important for understanding teaching as it is practiced than what can be explicitly formulated into reasons. To explore this particular issue is not to deny the others, or to somehow reduce the importance of reason. This could only be the case if the existence of reason was considered as completely unrelated to emotions.

Rom Harre (1986) points out that there has been a tendency among both philosophers and psychologists to abstract emotion as an entity—call it love, anger, and so on—and then try to study it. But what we have, he argues, are angry people, and grieving families and funerals in a concrete world of contexts and activities. With that caution in mind, Harre suggests that we rethink the question of what an emotion is, beginning with a study of how a culture *uses* its emotional vocabulary to pick out episodes in their emotional lives. Harre further argues that it is not simply the case that these descriptions should be deleted to reveal what an emotion *really* is. An emotion can be only how this or that person understands and uses the emotion in his or her culture. This process of studying emotion is the one I use in my work. It is what David Schwayder (1965) called *descriptive metaphysics*, namely, the process of describing at a metalevel what is underneath an activity as opposed to what must be.

Furthermore, I consider the traditional assumption that emotions are essentially private to be misguided, reductionist, and simplistic. Without ignoring the role of private experience in emotions, I suggest that the way we experience emotion is partly constructed by how we view emotion in our everyday discourse within the cultural context in which we live. Instead of assuming that there is an entity or an essence in teachers' emotions in science teaching (called frustration or excitement and so on), I begin by asking, How is frustration (or excitement), and other descriptions that cluster around it, actually used or experienced in science teaching? Looking at how emotions are used in the practice of science teaching allows for the possibility that science teachers may use particular emotions in different ways. Given such-and-such circumstances in the classroom, how does a particular teacher perform? What does he or she do or say? How do his or her emotional responses evoke or maintain certain acts or emotions? Unraveling the manifestations of emotions in science teaching will lead us deep into the usefulness of emotions in obtaining a better (empathetic) understanding of teaching science.

My concern for empathetic understanding is also related to the notion of *emotional understanding*. Denzin writes that emotional understanding "is an intersubjective process requiring that one person enter into the field of experience of another and experience for herself the same or similar experiences experienced by another. The subjective interpretation of another's emotional experience from one's own standpoint is central to emotional understanding. Shared and shareable emotionality lie at the core of what it means to understand and meaningfully enter into the emotional experiences of another" (1984, 137). It is important to acknowledge that in researching emotions emotional understanding may be useful at least on two levels. First, the descriptions about the emotional experiences that are involved in interpreting someone else's actions operate at a level where they truly evoke an emotional understanding of these experiences based on our own. In other words, emotional understanding is valuable in helping us recognize that what we see in a science classroom is enthusiasm, frustration, embarrassment, or shame; it helps us to reflect on the emotions being discussed and sense their pedagogical significance. Second, researching emotions encourages rich emotional understanding in science education more generally, that is, the recognition of seeing science teaching as an emotional practice that involves a great deal of (positive and negative) "emotional labor" (Hochschild 1983). The underlying assumption of both levels of emotional understanding that I refer to is based on the idea that we share common feelings and emotional experiences with others.

A commitment that remains constant throughout my research is the quest of what Lyotard (1984) called "paralogy," a quest that reaches out for the *new*, for new questions that will generate new inquiries. My rejection of any attempt to search for absolute certainties indicates my epistemological attitude—there is no room for certainties, but rather a struggle of *understanding* the difference, the diversity, the intuition, in other words the *unpresentable*. There is an anti-foundational spirit in paralogy, and its constructive power can provide refreshing ideas for science education pedagogy. I use the term "anti-foundational" to signal what Lather refers to as "the historical space in which we find ourselves 'after truth,' after certainties and absolute frames of reference" (1993, 688). I do so on the grounds that preoccupation with modernist assumptions embodies transcendental and teleological foundations that mystify the teaching of school science.

Such understanding of elementary school science teaching as the one I sketch here is *wholistic* (contextual and nonlinear) because it is interwoven in teachers' emotions, feelings, intuitions, perceptions and actions (see, for example, Du Bois's [1983] reference to a wholistic approach in knowing). Its purpose is to reveal the unseen and move through plural paths. My preference to paralogy consists not of apologetics but of an effort to demystify the foundational views of knowledge in science and in science education as they are expressed in the doctrines of coherence, continuity, and teleology. Demystification, in Lyotard's view, is an endless task. Yet, disrupting pervasive myths in science education has never been easy. I consider the attempt to create a *wholistic* understanding of elementary school science teaching as a worthwhile project because I strongly believe that the attempt to equate empirical reality with the whole of reality at the exclusion of tacit knowing has brought us to the point where we stand today: the unfortunate split between cognition and emotion (and so many other forced dichotomies), and the idea of rationality as the ultimate organizer of the whole of human life.

Obviously, my views tie well into the feminist argument in science education: if we make any attempts to reclaim a more wholistic science, then we need to make attempts to reclaim more wholistic ways to understand what goes on in science classrooms. Although noone has examined emotions in science education using a feminist lens, I think it is useful to consider how feminist ideas enrich this research, especially if we look at the feminist accounts of emotion in other areas.

In Sue Campbell's (1994) view, feminist analyses of emotion address a common concern, the dismissability of women. Feminist thinkers (e.g., Bartky 1990; Boler 1999; Campbell 1994; Code 1996; Game and Metcalfe 1996; Greenspan 1988; Griffiths 1988; Lutz and Abu-Lughod 1990; Laslett 1990; Rosaldo 1984) reject the dichotomy between emotion and cognition and, consequently, the dogmatic position that emotional response is irrational. Feminist writing challenges the idea that emotions, feelings, and bodies are in opposition to cognition, rationality, and the mind. Feminism (although not monolithic) questions the political motivation behind such dichotomies and the hierarchical control they imply.

Around this central theme (the dismissability of women) there are various other concerns that address different perspectives of particular types of dismissability. For example, some feminists, says Campbell, attacked the myths that our emotional experience is private and that it is (epistemically or otherwise) privileged. "These myths," adds Campbell, "encourage the false view that we could maintain a well-defined emotional life independent of the power of others to interpret our expressive behavior" (1994, 47–48). On this view, the feminist interest focuses on the authority of others to interpret the expressions of one's emotions. An examination of the association of the feminine with feeling, for instance, has been a traditional and historical ground on which to dismiss women. The association of emotions with nature or with femininity and the fear that they are both uncontrollable and immensely powerful justify patriarchical concerns that such inappropriate behavior needs to be controlled and not allowed to enter the public sphere (see, for example, Lutz and Abu-Lughod 1990). Campbell argues that a rational-irrational evaluation of emotions is inadequate because emotional expressions need to be understood in the sociohistorical context in which they occur. Inevitably, then, issues of power and authority cannot be ignored. "As feelings are formed through expression, people can exercise restrictive control over our feelings through controlling our acts of expression and thus dismiss or diminish the possibilities for finding or creating significance in our lives" (Campbell 1994, 55). As a consequence, explains Boler (1998), feminist theories and practices have systematically politicized emotions as a site of social resistance and transformation. Boler defines "feminist politics of emotions" as "the explicit analysis and resulting individual or collective actions, that challenge the historical and cultural emotional rules which serve to maintain capitalism and patriarchical hierarchy, particularly with respect to the arbitrary gendered division of public and private spheres" (1998, 49–50). A feminist politics of emotions encourages women to articulate their emotions and to develop alternative emotional expressions that challenge patriarchical ideologies.

To summarize the ideas that inform my approach in studying emotions, my approach has a nomadic character in the sense that it draws methodologies and theories from multiple areas, such as poststructuralism, feminism, philosophy, psychology, anthropology, and cultural studies. But how do all these ideas come together in the context of science teaching? In the last part of this chapter, I will undertake this task through the use of two vignettes from my research with Catherine.

The Birth *of Emotional Genealogies of Teaching*

> Your roots are full of viruses of transcendental terms. Your genes fall southward to the low last edge of China, lonely and broken. Your dreams are touched not of time. You desperately try to plant them everywhere. Genes, dreams. Transcendental China in novel phrases.

An adequate account of emotion has to take into consideration the context in which a teacher's emotions come into being: the classroom. At the same time, it needs to take account of the history of the teacher; hence, snapshots of a few instances in one's science teaching are insufficient to give any depth of understanding. Exploring one's emotions cannot be done over a short period of time. The account I outline here is based on my understanding of the role of emotions in Catherine's classroom life for three years. This justifies my rationale that an ahistorical snapshot framework is essentially distorting. I describe my approach in terms of what I call *emotional genealogies of teaching*. As Foucault argues, "[Genealogy] must record the singularity of events outside of any monstrous finality; it must seek them in the most unpromising places, in what we tend to feel is without history—in sentiments, love, conscience, instincts; it must be sensitive to their recurrence, not in order to trace the gradual curve of their evolution, but to isolate the difference scenes where they engage in different roles. Finally, genealogy must define even those instances where they are absent, the moment where they remained unrealized" (1984, 76).

I define the concept of *emotional genealogies* as referring to an account of the events, objects, persons, and their relationships that are present or absent in the realization of emotions, and the ways that these emotions are experienced in relation to the self, the others, and the world in general. Applied to my work, an emotional genealogy of Catherine's science teaching seeks not to trace the gradual evolution of her emotions but to record the singularity of various events that make some emotions present and others absent. My approach is genealogical in the sense that I am interested in constructing histories of emotions through the experiences of an elementary school teacher. The specific focus of my research is how certain emotions are constructed in the science classroom and how they are transformed over the years (as mediated by values, philosophies, beliefs, and so on). Catherine's emotions in science teaching are a "history of the present" (Popkewitz and Brennan 1998), a history of her emotions' "presences and absences" (Grumet 1988) in her daily interactions with her students, parents, and administrators in the context of the science classroom.

In the following few pages, I share two vignettes from Catherine's science teaching; they show two things: first, the importance of the role of her emotions in her science teaching and, second, the value of constructing emotional genealogies of teaching in exploring a teacher's emotions. In the first vignette, I explore positive emotions—how Catherine constructs her excitement and enthusiasm about science and their importance for her teaching. In the second vignette, I describe a series of negative emotions—frustration, guilt, and lack of self-confidence for science teaching—and their development throughout Catherine's career. (Obviously, the limited space does not allow me to explore these vignettes in details; for more information on these and other vignettes see Zembylas [2000].)

Vignette One: The importance of excitement in science teaching.

"I feel the excitement and the richness of working with children when some of my younger parts are involved, enjoying the creative nature or the fantasy nature of our in-depth explorations," said Catherine during one of our discussions about how her excitement affects her teaching. In part, this was very much related to her willingness to be more reflective about her ways of feeling. She admitted that she felt more comfortable doing this now than earlier in her career. She even discovered that it was more interesting to reflect on her emotions and that it helped her to analyze what she was doing: "It involves analyzing how I approach things, how I think about process and information. And I don't know how they are going to affect us [in the classroom] or my teaching, but I would assume that they are at least in part responsible for my willingness to think about things, to feel something, or to talk about things. And thinking about and expressing how I feel, 'Well, I wonder' or "I am excited about [so and so.]'"

Catherine became more excited and interested in things when she was motivated to find out more or when she saw her students being excited. For example, around the middle of October in 1998, Catherine and her second-grade students were exploring the growth of triops. There was a certain degree of mystery and excitement about observing the triops, since this was their second trial to hatch triops after the first one failed earlier in that month. At the beginning of the year the children examined the hatching of chicks; therefore, on this particular day they began a discussion about the differences between hatching chicks and hatching triops. This was not something that Catherine had intended to do, as she explained later, but since one child suggested that "the triops grow quicker than the chicks," more children wanted to share their ideas about the differences between the triops and chicks. An exciting atmosphere began gradually to emerge with children trying to find as many differences as they could and Catherine documenting their ideas on the board. Then the children began to ask questions such as whether it was safe to touch the triops and what precautions they thought they should take to handle the triops. Catherine seemed obviously excited (as she reflected on this episode after the class, and as she wrote in her emotion diary) and decided to provide them with as much time as they wanted to ask more questions and discuss differences between chicks and triops. After the episode, I asked her to describe the feeling of excitement and to analyze how she became aware of it and the ways that it affected her pedagogical decisions, such as the decision to allow more time for discussion and exploration. "I'm much more aware of the excitement, and yet again when it happens, I'm trying to understand why was that so exciting? What really made that experience so wonderful. I still go back and reflect further and frequently for me it is partly, in my experience, how it was similar to or different from something that happened to me when I was younger or last week—as well as how fascinating it was to see kids being engaged, so I was engaged, too."

Similarly, when the children were engaged in creating a time machine to travel through time, Catherine expressed her excitement about the children's ideas and how this excitement influenced her perspectives about the value of the activity. "I was very excited about all the things that students have come up with. I watch the class, my children, become very intrigued about the time machine in a different way. We watched a little bit of *Back to the Future* and they were fascinated with the car and how things worked. And they really got into the lever and started talking about a lever that goes up and down and how they could build something similar for their own time machine. So, I was fascinated watching their excitement and seeing things that triggered that." Catherine described how amazed she was to watch her students being so immersed in this activity. She also explained how their feeling of excitement kept their interest and gradually changed their perspectives and made their questions more focused, since they became more purposeful at what they wanted to do.

> It was amazing to watch. I just started kind of playing, saying, "I'm gonna work over here." Some kids wanted to work with me, a lot of kids just went over and started doing things. And then every once in a while they'd come and ask for a job. And some made up stuff and it was fascinating to watch how involved they got, and how quickly the jobs went from something that you just plug something in to something that they were spending a fair amount of time doing something before they came back. And they asked all kinds of questions. I of course spent a lot of the time just in absolute amazement. Then we started listing some of the parts that they needed to make, and they seemed much more purposeful about what they were doing and what they wanted to do.

Catherine was delighted that the children were engaged in some high levels of thinking. She believed that the fact that she provided space to her students to change things and have control over what they wanted to do by making the process meaningful and fascinating to them was one of the reasons that the children's excitement kept growing. Seeing her students being excited affected not only her emotions but her future decisions and actions. "I was excited. I thought it was a really exciting process. My next piece is I need to make sure that each of those kids works with some other people and then go back and talk about their inventions to everyone. They need to share this excitement." As Catherine said elsewhere during our discussions, her excitement seemed to nurture the children's fascination and vice versa, depending on the activity and the circumstances. But there was a mutual support of one another that created and sustained their positive emotions and attitudes.

Vignette Two: The development of self-confidence in science teaching

In Catherine's early years of science teaching a sense of guilt as well as deeply felt disappointment derived from her feeling that she was not being as successful as she aspired to be. Initially she was blaming only herself, but at some point she realized that there were many involved in making things the way they were. She specifically referred to the system's philosophy—for instance, the emphasis on science content knowledge and standardized testing as well as the masculinized character of science itself as objective and unbiased—and to other colleagues within the school where she worked. The poor communication among school staff was a serious source of frustration because she felt that the effectiveness of her own science teaching was obstructed by other people saying that she was different and that she wanted to try dangerous ideas in contradiction to the school's emphasis (as they perceived it) on science content knowledge. Other teachers felt that Catherine's students would lack knowledge, and they questioned her values and ideas of providing opportunities to her students to explore their own ideas. Catherine felt that she was dismissed; this feeling of dismissability hit at the heart of Catherine's self-confidence and self esteem and led her to question her teaching approach and her values.

The turning point in Catherine's career occurred when she taught kindergarten fifteen years ago. It was then that she saw for the first time how she and her young students could articulate their own views as *felt* not merely as *thought*. "When I went to kindergarten," says Catherine, "I felt comfortable for the first time talking about feelings. I felt I didn't need to know all the answers. Also, going to kindergarten I recognized the incredible need that everything had to be hands-on. Feeling comfortable for the first time to say to my students, 'I don't have all the answers,' opened the doors for a deeper reflection on my feelings, my teaching, and my self-confidence." She still remembers a young black boy coming to her one day and saying, "Ms. Myers, I found this wonderful rock in my backyard. This is the prettiest rock I've ever seen. I want to learn more about my rock." Since that day, Catherine admits that she had to give up using *reason* as the only way of organizing an activity, of experiencing *knowing*, or of reflecting on her teaching. She began to see the countless instances of emotionality in herself and her students: the feelings of success and failure at finishing a science project; the students' curiosity at her announcement of a new science investigation; her frustration when a science conversation with her students seemed to lead nowhere; her own personal passion and interest at exploring sea life and space.

"I remember curiosity as a kid," says Catherine, "being fascinated just by things. And I think for me some of what happens with children is just exploring stuff and thinking about it and wondering with them as if I am a kindergartner or a first grader or a second grader. And I think my guess is that all of us have had those experiences when we were very young, about being really mesmerized by watching a worm, or a beautiful, colorful rock. And it seems very much so that a place to begin in science education is targeting what students or a teacher remembers or discovers as being interesting or being excited about."

Catherine now feels that the need to emphasize the importance of feelings in science is a valuable way of knowing the world. "Emotions," she says, "are ways with which we know the world around us." Catherine uses multiple ways and audiences to articulate her beliefs, thoughts, and emotions. She keeps a self-reflective journal (one part of which is an "emotional diary") and she often shares various excerpts from it with her students, excerpts about things she wonders, things she is excited to learn, or things that frustrate her. She frequently tells the students that our feelings sometimes tell us many things about what we value, and that we can use this information to make better decisions. How she and her students feel sometimes determines the direction of their science investigations: if there is frustration and confusion, she tries to alleviate that by infusing activities that create positive feelings and bring back the lost enthusiasm; if there is excitement, she encourages her students to express that in many different ways—role plays, poetry, art, design, music, and dance. Catherine often shares with her students how much she loves science, and she uses her emotions to make science more exciting and refreshing for herself and her students.

The development of her propensity for self-reflection has been fundamental in Catherine's efforts to overcome her feelings of shame, guilt, and inadequacy in science teaching. She is constantly engaged in self-reflection—through writing a journal, sharing her ideas with colleagues and university researchers, attending conferences, and presenting at workshops. By reflection Catherine means that she tries to become deeply aware of her strengths and weaknesses, her positive and negative emotions, and the ways that those influence her science teaching.

As a result of constantly looking for ways to understand her teaching and herself, Catherine points out that over the years she has increased her self-confidence. "I don't think I had a broad enough understanding of what science was back when I was a young teacher," she said once, and then added: "Maybe it was when I freed up to 'Oh, I see that science is everything I do! OK!'. Probably the biggest thing was just self-confidence. And in writing and reading more about ways to approach looking at science and thinking about my ideas and sharing my feelings with my students and some of my colleagues. I think all of those experiences with things that made me feel more comfortable and more confident about 'Gee, it's not that I know a great deal more than I knew ten years ago about individual topics,' maybe this is about thinking more about what I think and how I feel it's important." Catherine gradually regained her self esteem by changing her philosophy about what science meant to her: her idea that science was not a collection of facts but a fascinating way of knowing a way that never ended. As she pointed out, " I don't need others to validate what I am doing any more. My students' satisfaction and growth and my own excitement at what we are doing are enough."

Theorizing about the Emotional Genealogies of Teaching

The two vignettes that I have shared allow me to argue, among other things, that Catherine's emotions are not internally located within her individuality as a private experience or simply a biological phenomenon, but rather her emotions are embedded in social, historical, and political (ideological) contexts. In making this important distinction I am not saying that a framework of private consciousness or one of internalized ideology are adequate by themselves to analyze emotions. Campbell (1994) discusses the dismissal of emotions and argues that they are *collaboratively* formed. Therefore, I want to argue that although emotions are one's own, they are not formed independently of the rest of the people interacting with us: Catherine teaches science in specific historical, social, and geographical circumstances; hence, the classroom and school context interact with the physical ground of her being. What I suggest is that the emotions of a teacher need to be understood in terms of one's history and in the terms of the social context in which she or he teaches.

Teachers' emotions in elementary school science classrooms can be described in terms of at least the following three domains: how teachers express their emotions; what kinds of emotions they usually express; and how they talk about them, and how they use their emotions to plan, create, and motivate action in their classrooms. Several central questions may arise from this account on the role of emotions in science teaching: First, concerning the content of science teachers' emotions, what does it take for an emotion to have the kind of content that makes it an emotion? What elements/components constitute an emotion? What contexts create these elements/components that constitute emotions? Second, an epistemological question is, What makes such emotions constitutive of knowledge of some worth in science teaching, and of what worth? In other words, how are emotions, teaching, and science intertwined in the construction of knowledge and teaching of science? What forms do emotions take in particular contexts in the elementary science classroom? How does the teaching of science enable and constrain certain emotions of teachers? Also, by looking at how a teacher's emotions are constructed in a science classroom, one can explore issues that are relevant to the context of emotional justification, issues such as, How are a teacher's emotions in a science classroom justified by the situations in which they arise, and how do these emotions function in the justification of the teacher's actions?

Many questions are asked. I don't know if there are any satisfying *answers* to those questions or even if they can be pinned down. I will not even anticipate what they *should* or *could* be. Such questions require us to consider afresh in *every* classroom situation just how emotions have some significance, serve any function, and so on. These issues cannot be easily settled in any prescriptive terms. It is by *describing* teachers' actions and emotions, I think, that we may identify and understand the significance of emotions. Perhaps elementary school science teaching is influenced more by social practices, teachers' emotions, matters of temperament, and other nonrational factors than by rational arguments and grand claims about necessary and sufficient standards in science teaching. Nevertheless, it seems to me that the search for necessary and sufficient conditions or standards is a project doomed to failure because it fails to capture the imaginations, desires, and the tacit dimensions of human judgment.

Unfortunately, our traditional questions in doing research in science teacher education have been, *What* topics should we teach? *How* should we teach them more effectively? *Why* should we teach those topics and not others? And how we can ensure that certain professional *standards* are met? Setting standards of what teachers should know and be able to do, as many educators would point out, can certainly help ensure the profession against truly awful teaching. However, we seldom, if ever, ask the question, Who is the self that teaches science? to paraphrase Palmer's (1998) question. Or like Hamlet, I ask the question, Who's there? The "whats," the "hows," and the "whys" are all significant and important and yield valuable insights when we explore our questions and concerns in science teacher education. But none of them provides the insights that can come from asking the "who" question. There are obviously several components to this question—intellectual, emotional, spiritual—and they are all dependent on one another and interwoven in the human self and in education (Palmer 1998).

My vision in researching emotions in science education is to take the first steps in developing science pedagogies that invite teachers' (and children's) emotions into reflection. I hope that this effort will help construct a discourse that encourages a critical inquiry of the role of emotions in science education. I realize that this is not an easy task. Educators who share their emotions in the classroom or with colleagues and who invite students to do the same face serious challenges. Megan Boler refers to some of these: "How do we challenge one another's claims to experience? In an era of identity politics, how can we respect differences as well as develop collective and self-reflective skills of critical inquiry regarding the nature of our experiences and claims to truth? Does the invocation of experience and feelings necessarily lead to essentialist claims?" (1999, 111). I think these questions need to be examined more seriously; these and other related challenges deal with the connections of a teacher's emotions as central components of his or her lived experiences and the development of his or her capacity for self-reflection.

I know that some might criticize my approach of emotional genealogies of teaching as ambiguous, unclear, and unsatisfactory, saying that it inflicts a rhetoric that may be misleading. However, a fairer and more generous "reading" of this work may show some ways to a more open research approach, and to multiple understandings of science teaching as both a rational *and* an emotional activity. What I suggest here is that we need to transform both our understanding and our experience of science teaching. As a result of this shift of orientation we might be able to understand how emotions arise, develop, and become fertile in science teaching, as well as ways in which they may degenerate and the roles that these ways play in the "effective" teaching of science. I think that despite the difficulties concerning the description of these "emotions" (and the meaning of this term), an important truth emerges that may need to be preserved: many more questions about the role of emotions in science teaching come into prominence as a result of interrogating the process of how we teach science and the respective complexities of this process.

A final warning concerning my theorizing about emotional genealogies of teaching: Although one of the "language games" that I use to analyze the emotion discourses is an epistemological one, the analysis of the relationship between epistemological presuppositions in science teaching and the emotions can be very problematic, not least because traditional epistemology is often privileged while the emotions in science teaching have not yet been described in a socioculturally sensitive manner. I do not claim that paying attention to teachers' emotions alone merits epistemological attention nor that it alone can provide any explanatory accounts. Designating emotions as the place of interrogation rejects the hegemony of a narrowly conceived process of researching teachers' thinking and decision making as the arbiter of what counts as knowledge in science teaching, and of professional researchers as paradigmatically worthy knowers. This does not amount to a naive refusal to acknowledge the importance of epistemological questions, but it invites a call to address critically manifestations that may not be (only) epistemological. Besides, there is always the possibility, as Code (1996) suggests, that epistemology misdescribes all but a select part of human activity.

Conclusion

> Love is simple. Yet I shall leave nothing behind me but unfinished poems mixed with thick leaves. [I feel the rhythm of a snail. Its body relativizes space: Neither up nor down, neither left nor right are meaningful as long as it meets new flowers in its path.] The wind lifts silence and my words. The poem still remembers that it is blind.

When we pay attention to thoughts and emotions that science evokes to us, we may rediscover a fascination and a wonderment in how we teach science. Noone can deny the overwhelming emotional complexity of science teaching. Teaching science cannot be done well by disembodied intellects but must be done by whole persons with emotions and intuitions and minds as well. This does not mean that one has to deny professionalism by teaching in ways that nurture one's emotional responses as part of whom she or he is. It is interesting to observe that the most powerful (and traditional at the same time) notion of professionalism in teaching has been modeled after male professionals in medicine and law, professions which require them to maintain a professional distance from their clients and avoid any emotional involvement with their client's problems (Grumet 1988). Grumet speaks for women educators and warns about the dangers of adopting this patriarchical model of pedagogy: "When we attempt to rectify our humiliating situation by emulating the protectionism and elitism of the other 'professions,' we subscribe to patriarchy's contempt for the familiar, for the personal, for us" (1988, 58). Developing an awareness of our emotional responses as one of our many ways of knowing, we can sort our experiences, our anxieties, our fears, our excitements and learn how to use them to appreciate how we can teach science in refreshing ways. Experiencing and reflecting on our emotions as educators, we can constructively integrate them with our teaching of science.

To conclude, let me summarize from a slightly different perspective. Although it is likely that an opposition between reason and emotion in science teaching will continue, I hope that a new pattern will emerge: one that seriously interrogates the unacknowledged assumptions of some dominant dichotomies. The theories of emotions, theories that have dominated Western philosophy, have assumed that emotions are disturbances; as such, emotions have been considered as distortions of truth and rationality. I do not expect that such views will change any time soon. My goal has been to show the problematics of these dichotomies in science teaching and to justify my central thesis that we need a more *wholistic* (contextual) approach in our efforts to understand the happenings of science teaching. I regard my work (and especially my study with Catherine) as a preliminary sketch of the role of emotions in science teaching; this would require considerable development (far beyond the constraints of a single essay) to begin even to be somewhat convincing. My aim is description, not salience to all the emotional aspects of elementary school science teaching. Obviously, there is much that remains to be done in this area, and what I offer here is simply an example.

References

Abu-Lughod, L., and C. A. Lutz. 1990. Introduction: Emotion, discourse, and the politics of everyday life. In ed. C. Lutz and L. Abu-Lughod, *Language and the politics of emotion*, 1–23. Cambridge: Cambridge University Press.

Bartky, S. 1990. *Femininity and domination*. New York: Routledge.

Boler, M. 1997. License to feel: Teaching in the context of war(s). In ed. A. Cretkovich and D. Kellner, *Articulating the global and the local*, 226–243. Boulder, CO: Westview Press.

———. 1998. Towards a politics of emotion: Bridging the chasm between theory and practice. *APA Newsletters* 98 (1): 49–54.

———. 1999. *Feeling power: Emotions and education*. New York: Routledge.

Braidotti, R. 1994. *Nomadic subjects: Embodiment and sexual difference in contemporary feminist theory*. New York: Columbia University Press.

Campbell, S. 1994. Being dismissed: The politics of emotional expression. *Hypatia* 9 (3): 46–65.

Code, L. 1991. *What can she know? Feminist theory and the construction of knowledge*. Ithaca, NY: Cornell University Press.

———. 1996. What is natural about epistemology naturalized? *American Philosophical Quarterly* 33 (1): 1–22.

DeBoer, G. E. 1991. *A history of ideas in science education*. New York: Teachers College Press.

Deleuze, G., and F. Guattari. 1987. *A thousand plateaus: Capitalism and schizophrenia*. Minneapolis: University of Minnesota Press.

Denzin, N. K. 1984. *On understanding emotion*. San Francisco: Jossey-Bass Publishers.

Didion, J. 1979. *The white album*. New York: Simon and Schuster.

Dilthey, W. 1958. *Gesammelte Schriften*. Leipzig: B.G. Teubner.

Du Bois, B. 1983. Passionate scholarship: Notes on values, knowing, and method in social science. In ed. G. Bowles and R. Duell-Klein, *Theories of women's studies*, 105-16. Boston: Routledge and Kegan Paul.

Foucault, M. 1983. On the genealogy of ethics: An overview of work in progress. In ed. H. L. Dreyfus and P. Rabinow, *Michel Foucault: Beyond structuralism and hermeneutics*, 2nd ed., 229-52. Chicago: University of Chicago Press.

———. 1984. Nietzsche, genealogy, history. In ed. P. Rabinow, *The Foucault reader*, 76-100. New York: Pantheon Books.

Franks, D., and V. Gecas. 1992. Current issues in emotion studies and introduction to chapters. In ed. D. Franks and V. Gecas, *Social perspectives on emotion*, Vol. 1, 3-24. Greenwich, CT: Jai Press.

Gadamer, H-G. 1975. *Truth and method*. Trans. G. Barden and J. Cumming, New York: Seabury Press.

Game, A., and A. Metcalfe. 1996. *Passionate sociology*. London: Sage.

Garrison, J. 1997. *Dewey and eros: Wisdom and desire in the art of teaching*. New York: Teachers College Press.

Greenspan, P. S. 1988. *Emotions and reasons: An inquiry into emotional justification*. New York: Routledge.

Griffiths, M. 1988. Feminism, feelings, and philosophy. In ed. M. Griffiths and M. Whitford, *Feminist perspectives in philosophy*, 131-51. Bloomington: Indiana University Press.

Grumet, M. 1988. *Bitter milk: Women and teaching*. Amherst: University of Massachusetts Press.

Harre, R., ed. 1986. *The social construction of emotions*. New York: Basil Blackwell.

Hesse, H. 1961. *Steppenwolf*. New York: Holt, Rinehart, and Winston.

Hochschild, A. R. 1983. *The managed heart: Commercialization of human feeling*. Berkeley: University of California Press.

Jaggar, A. 1989. Love and knowledge: Emotion in feminist epistemology. In ed. A. Garry and M. Pearsall, *Women, knowledge, and reality: Explorations in feminist philosophy*, 129-55. Boston: Unwin Hyman Ltd.

Johnson, M. 1990. *The body in the mind.* Chicago: University of Chicago Press.

Keller, E. F. 1983. *A feeling for the organism: The life and work of Barbara McClintock.* New York: W. H. Freeman and Company.

Laslett, B. 1990. Unfeeling knowledge: Emotion and objectivity in the history of sociology. *Sociological Forum* 5: 413–33.

Lather, P. 1993. Fertile obsession: Validity after poststrusturalism. *Sociological Quarterly* 34: 673–94.

Latour, B. 1988. *The pasteurization of France.* Cambridge, MA: Harvard University Press.

Lutz, C., and L. Abu-Lughod, eds. 1990. *Language and the politics of emotion.* Cambridge: Cambridge University Press.

Lyotard, J-F. 1984. *The postmodern condition: A report on knowledge.* Trans. G. Bennington and B. Massumi, Minneapolis: University of Minnesota Press.

Nias, J. 1996. Thinking about feeling: The emotions in teaching. *Cambridge Journal of Education* 26: 293–306.

Palmer, P. J. 1998. *The courage to teach: Exploring the inner landscape of a teacher's life.* San Francisco: Jossey-Bass Publishers.

Polanyi, M. 1958. *Personal knowledge: Towards a post-critical philosophy.* Chicago: University of Chicago Press.

Popkewitz, T. 1998. Dewey, Vygotsky, and the social administration of the individual: Constructivist pedagogy as systems of ideas in historical spaces. *American Educational Research Journal* 35: 535–70.

Popkewitz, T. S., and M. Brennan, eds. 1998. *Foucault's challenge: Discourse, knowledge, and power in education.* New York: Teachers College Press.

Rosaldo, M. 1984. Toward an anthropology of self and feeling. In ed. R. Shweder and R. Levine, *Culture theory: Essays on mind, self, and emotion,* 137–57. Cambridge: Cambridge University Press.

Rosser, S. V. 1992. The gender equation. *Sciences* Sept.–Oct.: 46.

Schwayder, D. L. 1965. *The stratification of behavior: A system of definitions propounded and defended.* London: Routledge & K. Paul.

Toulmin, S. 1972. *Human understanding.* Princeton, NJ: Princeton University Press.van Dam, E., and J. Steutel, J. 1996. On emotion and rationality: A response to

Barrett. *Journal of Moral Education* 25: 395–400.

Weber, M. 1949. *The methodology of the social sciences.* Trans. E.A. Shils and H.A. Finch, Glencoe, IL: Free Press.

Wittgenstein, L. 1953. *Philosophical investigations.* Trans. G.E.M. Anscombe, Oxford: Basil Blackwell.

Woodward, K. 1991. Introduction. *Discourse: Journal for Theoretical Studies in Media and Culture* 13 (1): 3–11.

Woolf, V. 1931. *The waves.* New York: Harcourt, Brace and Company.

Zembylas, M. 2000. Emotions and elementary school science teaching: Postmodernism in practice. Ph. D. diss., University of Illinois at Urbana-Champaign.

Chapter 6

Visions of "Science for All" in the Elementary Classroom

Elaine Howes

I trace the source of my interest in science to my childhood explorations of the swamps and woods and abandoned farmland near my family's Midwestern home. Then, I didn't think of my relationship with nature as science; it was adventure, solitude, and solace. School was where science, in an admittedly attenuated form, existed for me. Of science in elementary school, I have one vague memory of a weather-related study. High school science consisted of one miserably dull biology course; the high point of this experience was my uncharacteristically seditious choice to walk out of the tenth-grade classroom rather than pith a frog. When I became a mother, my daughter accompanied me on my treks through mid-Michigan's meadows and swamps, which were quickly being overrun by the suburban sprawl that now completely carpets my childhood home. She was less intrigued by the details of the living things that surrounded her than by the story-making opportunities of private and overgrown places.

Even when I chose to study biology in college, partly due to its prestige, but mostly because I was fascinated with the living world, my own vision of science excluded any reference to the open-ended nonlinear activities of my own and my daughter's childhood. I left my physical and mystical relationship with nature behind to jump onto the molecular biology bandwagon. I concentrated on genetics, biochemistry, and the fascinating inner workings of the cell. I avoided ecology and the health sciences; I joined the vanguard of the post-World War II biology—a place where human society, ecological interactions, and even whole organisms were left behind in a search for understanding of the very small. My degree earned me a position in laboratory science, where I continued to use the tools of modern biology to peer into the minuscule worlds of cells and the biochemical reactions that made them run.

A few years of this left me yearning for a larger world that encompassed whole organisms and ecosystems, as well as labor more directly responsive to my desire to make the world a better place for humanity. So, as many women before and since, I left the laboratory for teaching. Now I find myself attracted to scientific endeavors that insist on keeping the connections between living (and sometimes nonliving) things intact. This feels like a return to the joy I found in engagement with the biological world in my childhood, in its living complexity, outside of a test tube; it is also greatly enlivened by my work with preservice elementary school teachers. As are mine, their recollections of *school* science are overwhelmingly negative. Yet, when given the chance, they not only recall but also conceive of experiences with the biological and physical world that reflect the connections, joy, and excitement that school science is more likely to neglect than nurture.

Ever since I entered science education, developing instruction that welcomed and nurtured all students' relationships with the natural world has been my primary commitment. I've seen this commitment among my own students. It is those features of our common effort that I will explore here. I consider this exploration feminist because it values what preservice students know, believe, practice, and wonder about, rather than comparing their knowledge and dispositions to a predetermined ideal. It is also feminist because I value connections (in fact, they are forefront) between learners and the natural world; between intellect, emotion, psyche, and corporeality; and between people. Also, the great majority of my students are women. As I teach and conduct research, I attempt to respond to the lived experiences of the particular women with whom I work.

In what follows, I will first explain my practice of teacher research in elementary science education. This is followed by a brief description of the science methods course I teach, and of the students involved in the course and in the research out of which this chapter has grown. Then I explain the two organizing concerns that inspired this analysis: namely, students' enthusiasm for hands on science instruction and their conviction that science is "everywhere." Under these umbrellas, the results of my investigation of these two concerns are laid out, with students' words as evidence of the subtlety and complexity of their thinking in these two areas. I end with my modified perceptions of these concerns in preservice elementary science education.

Teacher Research in Elementary Science Teacher Education

It is common to sigh over the lack of content knowledge, and/or the lack of comfort, of the great majority of elementary school teachers in regard to science. I believe that this perceived lack of content knowledge is influenced by the construction of elementary school teachers and the work that they do as being empty of rigorous intellectual activity (Paula Lane, personal communication, August 28, 1999; Walkerdine 1992). Elementary teachers and their work are believed to be anti-intellectual, apolitical, and in need of expert guidance. This perception is linked to ancient images of women, images that have defined them as instinctively associated with nature and with children. Developing an understanding of elementary preservice teachers as a group demands that I discard this image and replace it with something else; this "something else," however, cannot be its mere opposite. That would entail my complicity in maintaining a status quo that demeans the traditional work of women, work that requires and nurtures habits of connection and caring. This would not be a problem if I wanted science to proceed as currently envisioned: as objective, reductionist, asocial, and detached. Then, I would simply look for traits and knowledge that people bring to elementary science and that allow them to practice and portray this image. On the contrary, I want science, and our Western relationship with nature, to allow in, maybe even base itself in, connections between people and the rest of nature and connections between people, individually and culturally. I want the traits that have been associated with women, and thus tossed out of the traditional descriptions of Western science, to be brought back in. Therefore I pay attention to those traits, specifically those that indicate connection, as these students exhibit them.

It may be a truism that teacher research teaches one more about oneself than about one's students. In this, teacher research can reflect the demand that feminist research *must* involve self-study, a kind of intensification of the ethnographer's "self as instrument" (Stanley and Wise 1993). By embarking on projects designed to study my students' beliefs concerning science education, I put myself in a position to be continually examining my own. Questions, ideas, and dispositions that students bring to our conversations reverberate in my consciousness for years.

My Vision of Elementary Science Classrooms

As a teacher educator, I introduce preservice teachers to my own vision of what science and science education should look like. Inevitably, it looks like something in which I would be happy to participate, as a teacher or as a student. I often feel stuck, as do my students, between my conviction that science teaching should be based in children's own questions and explorations and the demands of standardized requirements. I am convinced that learning happens best when students' curiosity and passions about the world are addressed. The teacher's role in this situation is to help children clarify their questions and develop investigations to address them; introduction of scientific content knowledge is also part of the teacher's role. *My* picture of a classroom that supports my own vision is one in which there is a lot of talk; evidence of students' investigations, in the form of drawings and writing, decorates the classroom; and students share their questions, investigations, and theories on an ongoing basis. Because the curriculum is based in students' questions and interests, it is inevitable that it will often be based in their lives, and therefore firmly embedded in social concerns (Barton 1998). One way to find out what these curiosities and passions *are* is to listen to students.

This leads to a central point in my classroom vision: The instruction will contain regular and ongoing opportunities for children to express what they are thinking, what they care about, and what they are wondering. These expressions may be through science talks (Bloom 1998; Gallas, 1995), clinical interviews (Bell 1995; Driver 1989), drawings and presentations (Howes 1998; Krajcik, Czerniak, and Berger, 1999), and drama. Thus the teacher will do more listening than talking, and the children the reverse (Meier 1995). A technical interpretation of this ongoing student self-expression is that it gives the teacher multiple means of assessing children's learning. While this is not a negligible point, my goal is somewhat different. I want teachers to be fascinated with children's thinking about the natural world and use what they learn from children to shape science instruction and curriculum. In addition, a fundamental respect for children, as thinking beings, is consonant with a critical feminist perspective which aims to upend power hierarchies and place learners, rather than science, in the center of classroom life (Barton 1998; Hazlewood 1996).

Because I work directly with teacher education students, my ideas concerning what I want for them are more actualized than those that I want for *their* students. Their stories from their own teaching contexts help me to develop practical visions that are directly applicable to elementary classrooms. It is these stories and examples that provide a great deal of the data for my research. As my research life takes me out into real classrooms to work with practicing teachers, real examples of the kind of science teaching I support will become part of my pedagogical repertoire.

The Course: "Science in Childhood Education"

In my elementary science education course, I encourage students to make links between the social, the personal, and the scientific; to view these realms as tightly connected and, thus, ideally, remove the objectivist aura that surrounds science and alienates so many students. As students weave these ideas with their own theories about and experiences in science, they nearly universally express a desire to teach science in ways that are welcoming to all of their students. The development of pedagogies that achieve this aim then becomes the challenge. Thus, my teacher research with these preservice students is based in searching for and creating ways to actualize the ideals of science for all.

Many of the students who come to my science education course, most of them women, have internalized a version of themselves as just not good at science. I hope to assist my students in leaving behind their feelings of inadequacy when it comes to scientific practice and knowledge. Two core aspects of the course address this goal: the autobiographical, in which students recollect their own experiences in science in and out of school, and the critical, in which we try to get beyond the façade of egalitarianism that science, and indeed our society and our schools, present.

Autobiography in Elementary Science Education

I don't claim any originality in the utilization of autobiographical reflection; requiring students to recount their school experiences and ongoing teaching experiences is a common tactic in education courses. In fact, I'm lucky in that I meet most students early in their teacher education program, when being asked to talk about their lives and beliefs is still a novel experience. They share these experiences with each other and read accounts of other people's experiences in science (Pomeroy 1994; Roth 1995). I choose these readings to indicate relationships with science that deviate from the typical "I love science and am good at it and worked hard and thus succeeded" and bring in questions concerning the structures of the dominant scientific enterprise as well as personal, emotional, even mystical possibilities in science.

The power in the assignment, I believe, comes largely from discovering that others have had similar experiences in relation to science and science education. The isolation in feeling oneself just not good at science may be eased by recognizing that the majority of one's classmates have felt the same. This aspect of the course takes on the feel of feminist consciousness-raising, as students compare experiences, notice similarities and differences, and create theoretical explanations for these experiences (Frye

1992; Hartsock 1979). Occasionally, students express race, class, and even feminist critiques. But more commonly their reactions are those of relief, leading to a willingness to try science out again, with the belief that it should be a qualitatively different sort of experience than that which they have known.

Julia[1] recollects this autobiographical episode as therapeutic, and links it to her belief in her ability to teach science with children: "I think if I had not become aware of the fact that I was afraid of science and like some kind of psychotherapy where I admitted that, openly, and found others who agreed and then kind of got through that and moved on to, 'Okay, we knew we were afraid of science and we're not anymore. Now we realize that no one knows everything, now we realize that questioning is okay, now we realize that learning can really be constructed in that environment,' I was really confident going in to teach it to those kids."

Critique in Elementary Science Education
I aim to support students in developing a critique of science in Western society, with the related implications for science teaching. Part of this critique is a development of the understanding that *real* people do science. This is not to argue that anyone can do science, but to indicate that real people are embedded in inequitable structures that perpetuate unequal opportunities. Recognizing this reality may help one see the very real barriers that prevent anyone from succeeding in science. One assignment to support this critique requires students to choose a scientist who falls outside the white male group with which we are all familiar and "become" that scientist for a mock symposium held during class time. As students study people of color and women who have struggled to succeed in professional science, they encounter the barriers that have been put in the way of universal success. In addition, many students choose nontraditional scientists, such as Chinese traditional healers, Native American potters, and even religious figures—these individuals bring to the fore our definitions of what does and what does not fulfill our demands for scientific activity and knowledge.

This critique is tightly connected to the autobiographical work, in that I want students to see that it is not their own personal failings but the structures of the discipline and practices of science itself that have caused them to choose other paths for themselves, or have even actively discouraged their participation. The stereotypical visions of scientists are very strong, in the media and in texts, and are even reinforced by many students' experiences with geeky male (and sometimes female) science teachers. Understandably, many young women do not find these images attractive. Associated with these images are practices that in themselves are often alienating. For example, science students are instructed in the ways of the scientific method, which actively attempts to delete emotion and personality. Objects of study are taken out of the living and human worlds, and often taken apart in ways that deny rather than explore their interconnectedness. In order to create a different kind of science instruction, one that will be attractive both to elementary teachers and all of the children they teach, teacher education needs to introduce preservice students to critiques of science that help them realize that the structures and rules of science are socially constructed and, therefore, not inevitable.

A different approach is to persuade preservice teachers that science is not only important but also "fun," while encouraging them to set aside their negative impressions of science in order to convey a positive image to children. However, alienation from scientific practice and knowledge does not come from nowhere. On the contrary, it is a sensible reaction to a scientific culture that is not inaccurately perceived as asocial, controlling, overwhelmingly masculine, and even violent (Keller 1985; Merchant 1980). Thus, skipping over students' perceptions and experiences and expecting them to uncritically celebrate science is not only inefficient, but also irresponsible. In refusing to embrace the "Ain't science grand?" philosophy (Hubbard 1994), we can avoid masking the harm that scientific practice and knowledge can do and have done. We can also refuse to blame students for their perceived failures in science.

Feminist critiques of science note these aspects of the Western scientific tradition and argue that they have been created by white, middle- and upper-class European and North American men and thus often foreign to women and many men. My own perspectives are based in feminist critiques of science, critiques that note the distance of scientific practice and knowledge from everyday life and everyday questions, its supposed disinterestedness and objectivity, and its aura of asocial masculinity (Bleier 1986; Fausto-Sterling 1992; Keller 1985). Throughout the feminist literature, these attributes of Western science are repeated so often as to begin to seem obvious, and therefore suspect. However, I recognize these tendencies in my own inconsistent relationship with the ideology and the practice of science. And just as I begin to think that maybe I've gone overboard, that my feminist perspective is outdated and no longer powerful, I see undeniable evidence coming up in the images that students bring with them—images that show that that the sexist structures of science, as portrayed in schools and the ambient social world, are alive and well.

They continue to succeed in alienating many women from science and also in stifling and rejecting the curiosity of schoolchildren. Making the structures of scientific practice—or at least of its mythology—explicit is vital in supporting students' learning to develop both a new vision of and a positive relationship with science teaching.

Therefore an early step in this course is to help students externalize their images of scientists. On the first day of class, I ask students to draw pictures of scientists. When I explain the assignment, I ask students to think of a scientist, and then draw the first picture that comes to mind. I explain later that the reason I do this is to reveal the unattractive images that we are able to edit or beautify when given the time can remain as part of our gut reaction to the idea of science. Students are invariably surprised by the persistence of the stereotypical images they and their classmates hold of the practice of science: Garbed in white coats and strange hairdos, their white male scientists blow up things using test tubes and arcane substances in laboratory settings.

Another early activity consists in reading a collection of short quotes that address the question, What is science? The quotes represent textbook and standards' authors, mainstream scientists and Nobel prize winners, feminist critics of science, and elementary and high school students.[2] I intend for our discussion of these quotes to further what I hope becomes a critical analysis of images—our own and others—of science and scientists. One of these quotes is by Elizabeth Fee: "The voice of the scientific authority is like the male voice-over in commercials, a disembodied knowledge that cannot be questioned, whose author is inaccessible." During a recent conversation inspired by Fee's statement, a student, Leah, exclaimed, "If that's what science is, we don't want to be a part of it!" This collection of quotes was designed to encourage discussion; it is not by accident that particular statements inspire strong comment from students. In fact, Leah's comment was followed by a discussion in which students examined issues of scientific responsibility; its separateness, sometimes literal, as in the Manhattan Project, from the rest of society; and the implications of headlong research in cloning in the light of human morality. But my point here is that if students are able to see and name things that they do not like about science—aspects of its image, its power, and its indisputability— then I don't want to try to persuade them to think otherwise. I would rather resist the forces that educate teachers to "deliver their students to the patriarchy" (Grumet 1988) via a science that is uncritical and therefore a conservative rather than liberating enterprise. Instead, I would like to support them in envisioning a science of which they *can* be a part.

It is impossible to introduce preservice elementary students to the range and depth of feminist thought that informs my own thinking about Western science and science education. Therefore, in this course, I implicitly focus on the connections between autobiography and critique. Simply put, my main objective in this regard is that students come to believe that science is created by people who are social beings possessed of the creativity, passion, prejudice, and personal goals of other humans. A comprehension of science as a social construction can be very useful in opening it up to those who are alienated by scientists' apparent escape from the messiness of everyday life into a cold (and dull) world of unbiased intellectualism.

The Students and Research Participants

When I first came to my current teaching position, I was quite taken by the apparent willingness of students to engage in issues concerning inequity in science education. This may be greatly due to the fact that their teacher education program is located in a large urban center. Every day they are in contact with people who represent a great complexity of races, nationalities, cultures, classes, and sexual orientations. They see in their placements children whose inheritance is African, Asian, Mexican, Central and South American, Caribbean, and, recently, Eastern European (among others). They know that if they are going to teach in the city, at least in a public school, they will need to learn to teach many students who are very different from themselves, particularly in terms of culture and class. However, while most of these students are white women, a substantial number are Asian or Asian American and a few are African American. (The great majority of students in this program are from middle- or upper-class families.) The students at this private graduate institution have been academically successful, in a variety of public, private, and international settings. While a certain sense of elitism pervades the college, there is also a strong sense of social responsibility in the general student consciousness, at least among elementary preservice teachers.

Methodological and Analytical Perspective

I see a large piece of my role as a teacher educator as providing contexts within which students can "compare notes" about their experiences, philosophies, and questions regarding science teaching. That is one reason why my research has largely taken the form of "interview-conversations," in which self-selected students and I get together for an hour or two several times a semester to discuss what's on their minds. Usually, but not always, these interview-conversations are group affairs, with two to five students and myself attending. The participants in the interview-conversations that provide data for this chapter had completed the elementary science education course and so were no longer "my" students. While I usually began the interview-conversations with a question, the students quickly took this responsibility upon themselves and did most of the talking, questioning, and recounting of experiences. This dynamic was especially noticeable when they were involved in student teaching: It was during this time that the most interesting talk came up, as students applied their ideas to classroom teaching and saw what came of them.

After the end of the course, I sent a letter and a proposal to all of the students who had taken an elementary science methods course with me in the fall of 1997 and the spring of 1998. I provided them with three possibilities: no response, which indicated no desire to participate; "Option 1," indicating consent to my use of their course work for this research; and "Option 2," indicating consent to my use of their course work and their willingness to participate in ongoing interview-conversations during their student-teaching year. Sixty-seven letters were sent out; eleven students chose Option 1; eighteen students chose Option 2. Out of the eighteen who chose Option 2, thirteen ended up participating in interview-conversations. Due to their full schedules, participants' attendance at these conversations varied; in addition, some of the sessions were one-on-one with me, while others were conducted with two or more participants (up to five).

The investigation represented in this paper began with my desire to explore students' goals, perceptions, and actions in relation to hands-on science and the idea that science is everywhere. Thus I went into this study with a predetermined framework. Within this framework, I searched the data for references to these two themes. Because I was not looking for themes to emerge, but rather for an exploration of themes that I knew from teaching were widespread in students' discussions about science education, the data that I provide in this analysis is not evidentiary, but illustrative. The things that they say are unique, and yet consonant with these pervasive themes. The students who are directly cited in this paper are Kathleen, an African American woman; Barbara, Karen, Melissa, Michelle, and Julia, European American women; Jeffrey, a European American man, and Kevin, a Japanese American man.

I consciously and unconsciously deploy my own experiences in science and science education to interpret the tales my students tell me about their experiences in, perceptions of, and relationships to science. The perspectives I've developed highlight certain aspects of my students' conceptions and pass over others. In addition, the perspectives themselves continue to change. Old convictions get buffeted about and roughened around the edges. As I continue to change and learn, I see things in students' recollections, beliefs, and dreams for themselves and their students that contradict, complicate, and reinforce my own. For example, I have noticed that students regularly state that "science is everywhere," a statement that I regularly respond to with, "I think of science is something that *people* do. I think of nature as being everywhere, but science would not exist without people." Nonetheless, despite my direct instruction in this regard—in the shape of responses to students' writing, and in discussion in class—the vision of science as everywhere regularly holds throughout the semester. I have also noticed that preservice elementary teachers enthusiastically embrace the possibility of structuring their science instruction around hands-on activities. I encourage this propensity and yet am somewhat troubled, as are my students, by the ease with which intellectual engagement with scientific concepts may be avoided if instruction is all hands-on in the absence of "minds-on" (Roth 1989). In the light of preservice students' stated beliefs and desires for their own students' experiences with school science, I have reevaluated these concerns. The result is not the removal of these concerns from my thinking about teaching elementary science education. It is, instead, an investigation of the reasons that preservice students enthusiastically adopt hands-on instruction and an exploration of their insistence that science has escaped the laboratory and taken its rightful place in the everyday world.

Organizing Framework: "Hands-on Science" and "Science Is Everywhere"

I have been repeatedly struck by prospective elementary teachers' eagerness to engage in hands-on science teaching. They place their growing understanding of active learning within a broad context where children are in direct physical contact with the natural world, whether it be in laboratory situations working with batteries or outdoors interacting with living organisms, water, and wind. This is all possibly an artifact of the way I teach my course, in which reflection upon personal experience often brings to the fore a dissatisfaction with science education and a need for something different to take its place. As they experience and read about hands-on teaching and develop open-ended scientific inquiries rooted in their own questions, they choose to fill the space left by their rejection of their memories of didactic and unengaging science teaching with child-centered and intriguing instruction. Importantly, they want this hands-on activity to be based in children's own questions, interests, and choices. They recognize the difference between hands-on activities designed to fulfill imposed curricular objectives and that which is intended to address students' own questions; they set the kind of teaching they envision in contrast to memories of filling in little blanks on prefabricated lab reports.

In my daily practice as a teacher educator, my assumptions have been several: Preservice students want to teach hands-on science because it will keep the children occupied and entertained, thus creating fewer management problems. Preservice teachers want to teach hands-on science because it is fun—it is a relief from the dreariness of text and lecture-based instruction. And, as they learn about constructivist theories of learning, preservice students want to teach hands-on science because they come to believe that children learn through doing. I have also been tempted to think that preservice teachers are attracted to hands-on science teaching because it provides a barrier between themselves and science content with which they feel less than comfortable. If the children are busy doing, especially within cooperative groups, the teacher can remove herself or himself from the role of expert and thus her or his felt ignorance becomes less of a problem. Hands-on science activities may also be conducted in the absence of minds-on activities, again avoiding the risk of showing one's ignorance.

Viewed differently, these apparently wrong reasons for choosing hands-on science teaching become meritorious. For example, removing oneself as the expert is necessary to allow students to develop their own lines of thought, and to learn to think of themselves and each other as experts, absent the stifling right answers of traditional science teaching and texts (Bloom 1998; Gallas 1995). This step allows teachers to join their students in inquiry or, at the least, create an atmosphere in which saying "I don't know" isn't an anathema, for teacher or students. The unquestionable authority of science, as so often presented in school, can thus be replaced with a more accurate set of requirements for scientific activity and the construction of knowledge. Rather than memorizing facts and concepts, students and their teacher can develop their own inquiries and use accepted scientific knowledge where applicable. Open-ended, question-based inquiry often quickly leads into territory where not only the teacher, but professional scientists themselves do not have answers. This experience alone helps to create a portrayal of science as an ever-changing discipline that recognizes the tentativity of our knowledge about the natural world.

I am not referring here to the kinds of activities that are provided in kits or mandated curricula. These materials and activities may well provide grist for the mill of child-created inquiry, if used for that purpose. However, if used as proscribed, they often set up situations in which the teacher may not only avoid sharing her expertise and her own questions, but be discouraged from doing so. Thus hands-on science without the minds of *both* the children and the teacher engaged in questioning, puzzling, and developing explanations is merely science teaching as usual, where some children will learn and most won't.

In addition, students working in groups may develop a vision of scientific inquiry that is more in accord with the ways scientists work together, replacing pictures of scientists working in mythological isolation. Stereotypical images of scientists—the most famous being Albert Einstein—provide the misinformation that single men working alone develop transformative theories. Particularly in modern science, knowledge creation (as exemplified in multiply-authored scientific journal articles) is a group activity. Scientists work together within laboratories and field-based contexts; also, of course, they build upon, disagree with, and learn from each other's ideas. Nonetheless, modern science remains a very competitive field. Most children in classrooms benefit from an atmosphere of cooperation. This is one of the contradictions that arises when one considers the implications of children acting as scientists.

As I studied more carefully these preservice students' descriptions of hands-on, inquiry-based science teaching, I realized that their own understandings of the complexities of keeping in play students' questions, scientifically acceptable explanations, and other aspects of student and classroom life go beyond my initial simplistic criticisms. Hands-on, inquiry-based science teaching, if constructed in opposition to the deadly crunch of timed experiments and constrictive lab reports, may reflect a vision of science that accords more closely with feminist revisions of science itself, particularly in its stress on connection and bringing the whole person to the study. Often, the students with whom I work take science out of the classroom and picture children exploring beaches and forests, camping out, and even flying to the moon. Therefore, hands-on science goes beyond its literal meaning to encompass any and all experiences with the natural world. In this sense, I see these students' insistence on hands-on science overlapping with their insistence that science is everywhere. I believe that these inclinations toward including whole selves in explorations of the whole world provide a good starting place for inquiry-based teaching.

Unpacking Hands-on Science: The Power of Prospective Teachers' Visions

In this section, I will describe aspects of the visions of science that these preservice teachers create and emerging forms of science teaching about which they can be enthusiastic. As explained above, I have used the ideas of hands-on science teaching, and science is everywhere to organize my interpretations of these students' beliefs and practices. The quotes used in this account come from papers written for the elementary science course, from classroom discussions, and from interview-conversations that took place over the year during which they were student teaching. There are three subthemes included in my description of students' ideas about hands-on science instruction. The first of these, hands-on instruction and personal engagement with the physical world, involves students' conceptions of the role that physical interaction with the non-human world plays in children's learning. These physical interactions tend to include the whole person—physically, intellectually, and emotionally as opposed to only intellectually. The next subtheme, hands-on instruction and environmental science, describes students' experiences and beliefs that lead them to believe that hands-on science instruction is an important pedagogical tool for creating science instruction in which children can connect with the rest of nature. The final subtheme, hands-on instruction and the importance of choice, explores students' desire to create instruction from children's questions and interests. As do I, my students struggle with the challenges that placing instruction in children's lives brings up in terms of learning the required (and/or intriguing) concepts that professional scientists have constructed about the natural world.

While students do not state, explicitly, that science *is* nature, their talk about the place of science in the everyday world, especially the natural world, indicates this assumption. This conception is definitely troubling, and yet by delving into students' ideas in this area I have found some interesting possibilities for dissolving boundaries between self and the object of study, between disciplines, and between school and not-school.

Hands-on Instruction and Personal Engagement with the Physical World

Some students describe situations in which relationships develop between learners and the objects of study, or the materials that the learners' hands are on. Karen recollects a science fair project in which she and a friend built a periscope in the absence of expert guidance: "Instead of spending weeks doing research on what it is and how it works, we built our own periscope. . . Rather, the experience of 'experiencing' the tool, a piece of matter, another way of viewing, illustrated one approach to learning that was natural and simple and that I had probably used a thousand times before. By asking a different, more focused question, mainly, 'how does it work?' I taught myself about periscopes."

Karen's language illustrates a relationship with matter, and the creation of a tool from matter, that is somehow different from what would have happened had she merely followed directions for building a periscope, a distinction of which she is no doubt aware. Lacking explicit instructions, she needed to get involved with the materials and learn how she could shape them to create the desired result. The interaction between the matter and how a periscope works was evident; these were not separated out. While Karen did not impute consciousness nor life to the materials with which she worked, her interaction with the materials—her experiencing of them—hints at a kind of technological relationship between scientists and their materials that strives to comprehend those materials on their own terms. More importantly, in relation to teaching, she names this approach "natural and simple"; it was not muddied with instructions that were separate from her developing understanding and her impetus to create.

Julia explains her belief that children's learning in science cannot be squeezed into the short periods allotted to science in typical school settings. Having just been through a series of job interviews, Julia has become acutely aware of the time constraints that are put on children's learning in school settings: "You want to have kids have quality time, putting their hands on something and then making themselves one with it. You know, understanding what it is in a very personal way. That doesn't happen in 40 minutes. That happens in hours where they get to play around, and feel it out, and talk with someone about it."

Julia provides a description of an immersion in study as becoming *personally* engaging, possibly even ignoring the boundaries between self and other as a mode of study. Julia's conception, and Karen's, of hands-on learning is distinctly different from frenetic "making science fun" activities, which are more reflective of the television children's science show host Bill Nye's speedy delivery than the focused engagement that Julia wants her children to experience. Rather than engaging students through explosions and bright colors, Julia is trusting to children's ability to become intrigued and involved with the materials of study. This is not to say that Julia does not see the value of children's enjoyment of unusual and colorful scientific work; in fact, a great deal of Julia's instruction was based in what might be termed "exploratory chemistry," as she supported her students in developing and investigating combinations of everyday household materials. However, her students did not watch their teacher create pretty colors or strange concoctions; they did this themselves. Julia describes her experience of this aspect of her teaching, and her efforts to utilize children's fascination with their creations in helping them learn the recording processes of science: "A lot of them would come to me with these purple globs, you know, things that they wanted to do again! They wanted to do it at home; they wanted to do it in other classrooms. And I was like, 'Well, what'd you use?' We'd talk about how you document what you use, and then how you can alter it and compare it."

Julia also recognized the importance of human interaction—specifically, talk—as an avenue toward understanding. As quoted above, students need to talk with someone about what they're experiencing and exploring. She includes herself in this exploration; she is more comfortable questioning along with the children than providing answers, and she is pleased that her young charges learned to approach her more as a colleague than an expert: "Kids, after a while, got used to the fact that I didn't necessarily have the answer. They began to come to me more with observations, and sort of a collegial question, kind of like, 'Wow, I wonder why it does this.' The question wasn't 'give me the answer.'"

Julia's desire that children experience science as making themselves one with the materials of study is not something that textbook delineations of scientific method condone. Scientific method, as generally portrayed in school, argues *against* personal involvement and *for* strict boundaries between self and other and between intellect, emotion, and the physical world. In Julia's case, the children and their teacher are portrayed as engaged in a personal and communal interaction with materials. Within this context, Julia introduces the basic scientific process of recording what one does in order to duplicate it and share it with others.

Julia did not, before this course, consider herself a "science person." Barbara, however, was successful in science, particularly environmental science. Barbara provided an example of connections between intellect and emotion in her recollections of inspiring geology field trips. She described the passionate relationships with the objects of study (in this case, rocks) that her professors revealed. She interprets her professors' scientific activity as not only hands-on, but imbued with an excitement she calls "love": "Working with the geology department as much as I did, I saw science in a very different way, a very hands-on type of way. It really inspired me because [the professors] just *loved* rocks. . .And I'm like, 'I am *so* excited,' because they really loved what they did, and they really explained what they did, and they really made sure that you could get your hands on things, and that was what I loved so much about it."

Loving one's work and loving one's objects of study are not identical, and innumerable stories of scientists' passion about their work will convince neither teachers nor their students that science is for them (and vice versa). Nonetheless, Barbara found this sort of love inspiring, and indicative of excellent science teaching. In fact, she cited this experience with her geology professors as an impetus to study science in college. This aspect of Barbara's own science learning may have made her particularly eager to teach science; however, as described below, she was faced with teaching in a field with which she was not comfortable, and thus was put on equal footing with many of her elementary teacher colleagues.

Barbara's cooperating teacher welcomed her eagerness to teach science. While the topic was mandated (electricity), and specific materials supplied, Barbara was in charge of the instruction and the children's work during this extended unit of study. She indicated some trepidation about the content, because while she felt generally strong in her science knowledge, electricity was a topic with which she was not comfortable. Her initial concern quickly gave way to delight as she observed the children become energetically involved with the materials: "And once we got with the hands-on stuff, with the wires and the batteries and the light bulbs and the motors, it took on a whole life of its own . . . Like for instance [a student] diagnosed with speech and language [difficulties] put an index card on a motor, and [said], 'Oh wow, look!' And put a marker on it, and started making art, with the motor."

Barbara testified that she was taken by the ability of the materials of electricity to engage children who had been diagnosed with speech and language difficulties in such a way that they became intellectual leaders in the science classroom. This phenomenon of a more bodily science allowing so-labeled learning disabled students to shine comes up frequently. (For one clear example, see Tutt and Rosenthal 1995). Because hands-on work does not necessitate reading, writing, or speaking in a large group, children who struggle with these typical school activities can shine in hands-on contexts. This interpretation is especially apt if the work of the students is open-ended—since there are no lab reports to fill out, and no specific instruction to follow, children may explore and create utilizing skills with which they *are* adept (Gardner 1983).

Barbara is too sophisticated to believe that it was the materials alone that allowed this to happen. She also values the development of community in the classroom, and said "I wanted them to realize the most important thing in science is working in groups, and you have to be nice to each other, you have to respect one another, you can't be pulling at things, everyone has to have a chance." Later in the same conversation she described how she had introduced new scientific ideas to the students at the beginning of the electricity study, and was convinced that, although the students found this segment of instruction dull, their hands-on work would not have been so rich if she had not done so. Through this experience, Barbara seems to have reached at least a temporary resolution of a common question that preservice students are faced with: How and when can one introduce scientific content into hands-on explorative contexts? In a sense, she has rejected an inquiry model that places exploration *before* the introduction of explanatory concepts, a model that I myself utilized in the methods course. Instead, she introduced the scientific concepts *first* and was convinced that this made the children's hands-on experiences richer.

In my mind, Barbara's story brings up an ongoing issue in science teaching: It is not an easy thing, even if one is an expert in any given content, to know when to introduce expert knowledge into classroom inquiry. If introduced too early, it may shut down students' questions and interest; if introduced too late, students may become bored and even believe that there is some secret knowledge that they are not able to access due to personal failing. I do not believe that we can consider our students to be successful unless they have learned some of the basic concepts of modern Western science. While Barbara has provided evidence that her students are engaged in an exploration of electricity, I am unable to say what they have learned in terms of the standard explanations of the workings of electricity. In any case, I believe that Barbara is utilizing the hands-on aspects of this mandated curriculum to actualize her conviction that all children can succeed, even excel, in science.

I chose the above examples because they concern nonbiological science. A substantial number of their stories indicate that preservice students bring with them positive memories of interactions with physics, chemistry, and geology, memories that challenge the stereotype of elementary teachers as comfortable only with the life sciences. They provide a welcome relief from the age-old mindless merger of women and nature (as defined by masculinist philosophers, the biological sciences, and the popular media). However, these examples, while absent of any talk of avoiding bias or the importance of objectivity, instead argue for intimacy and personal involvement, even when the materials of study are not easy to anthropomorphize, as would be the case if the children were studying the living world. These students' talk in this regard is reminiscent of Keller's interpretation of Barbara McClintock's view of scientific inquiry in that they work to develop connections, rather than establish distance, between the learner and the objects of study (Keller 1983).

Hands-on Science and Environmental Science

Many students express a desire to teach a science that illuminates and reinforces human connection to the rest of the world. This desire is most evident when students are speculating about teaching in areas that are linked to environmentalism. As Barbara said, "The whole thing about environmentalism is that I think it shows children that we're not only connected to other things, but [that] there are relationships with other things, and other people and the whole world."

Other students voice their conviction that environmental science and/or environmental education are powerful contexts for themselves as science learners and will be intriguing for their own students. Melissa, like Barbara, held a degree in environmental science. When I asked her what she would do if the curriculum were in her hands, she responded, "The first thing that jumps out would be water pollution .I think that that's an important thing for kids to see and to understand; just the idea that water is not everlasting and it's not always clean and it's not always going to be there. Talk about how the fish are affected by not just what's in their water but what gets in their water, and where does it come from, what does pollution come from. But you can show, I don't know how, but, the clogging properties of oil and just what happens even if we did a big plant, or something, and poured oil on it and saw what happened."

Melissa is beginning to create actual hands-on, classroom experiences that will help her students understand the interconnectedness that environmental science represents. Kathleen, however, dreams of taking students outside in order to jump into the real thing. This is likely due to her experience as a Girl Scout camp counselor. She also brings environmental science into her argument: "I think experience is the key, giving them opportunities to do science. And, more than just talk about it, and more than like, "here's the frog, here's the scalpel, you should cut there." Because that's just not working. [At camp] we were talking about environmental science a lot, because one of my camp colleagues is a science teacher, and so we [had] great conversations about science. And we're talking about that rolling around in the mud thing. . . But we walked out there and it's like all this muck and stuff, and it's sort of like aren't there leeches in this water? And like what does it mean? And then our other kids are doing water testing."

Kathleen's vision of doing science is one that has children outdoors away from lab benches and canned experiments. In other conversations with Kathleen, she has stressed the importance of wonder in science teaching. She appears to recommend unstructured experiences with the natural world as the best way to involve children in scientific study. As they explore the wonders of the natural world, their questions can serve as a basis for further scientific exploration.

Kathleen's stories of her camp experiences range all over the subdisciplines of science, including earth science, biology, and physics. She described the way one of the camp counselors takes the girls on "edible hikes—and they find the blueberries and the raspberries and the mint leaves—that's science." Kathleen also utilizes her expertise with canoeing to support an argument that science is everywhere: "One day we had three canoes that were full of water. And we had to do this mad canoe rescue. It was serious only in that we're sitting here watching this, like physics at its worst, happen. Even when the kids, they can't row the boat because somehow you don't get both oars in the water at the same time, and most people are right-handed, so it always happens with the right hand. The boat's just spinning around. That's science."

In Kathleen's metaphor for science, whole *bodies* are involved. Here Kathleen is providing an image of science very different from a model that isolates scientific method from the physical, the sensual, and the social. Kathleen's young scientists are involved in learning about the world with their whole selves, and they're figuring out how to navigate it together. Thus Kathleen's vision of doing science spurns the partition of intellect from action and physicality. She has also strongly stated that the teacher should share her expert knowledge, and does not see this as a way to shut down kids' thinking. In other words, she does not appear to agonize, the way that I do, about when and how to introduce scientific knowledge into children's explorations. Because I know that Kathleen welcomes scientific facts and explanations into her own scientific experiences, I do not believe that when pushed, she would state that science and nature are the same thing. It is more as if she sees science as an enriching, although not necessarily primary, factor in children's interactions with nature.

When one is convinced that science is everywhere, opportunities for learning and doing science abound. They needn't be in a book, in a laboratory, or even in a school. Science experiences can also become more physical and more related to the natural world. The whole person becomes involved in learning; learning is complex and involves all aspects of the personality; it is no longer purely intellectual. Below, Michelle describes a class period during her teaching about leaves, in which she took first graders to a park. She describes their interaction with the wind and leaves, and her enjoyment in being able to "call it science": "We went to a park, and it almost seemed like [the students] had never really been out running around in the park before. And the wind was blowing that day and the leaves came down, and a few of them ran under the leaves and said, 'Look, it's a leaf shower!' It was just exciting to see their excitement, and then be able to call it science." Michelle's story reminds me that learning about the world doesn't have to be something that is controlled—it can be a spontaneous and joyful experience. In fact, Michelle decides to call it science; it is not science as defined by a required curriculum or a standards document.

During one of our interview-conversations, Barbara spoke at length about her choice of environmental science in college because it encompassed many things that interested her, namely, geology, writing, and political science. Michelle, at this point, noted, "[In] high school, it was all based around Regents, we had nothing environmental; as a matter of fact, environmental science is something I would be more interested in than in any other." Michelle, here, brings up a point that continues to put up barriers to full-out inquiry teaching, in which environmental studies could play such a powerful role: There are tests to be passed, which designate particular content to be taught. As with the matter of student choice, the role of standardized tests is a powerful conservative force.

Hands-on Science and the Importance of Choice

These preservice students' focus on hands-on work as a way to help forge connections is also evident in their desire that they will be able to develop instruction in which children will be allowed and encouraged to pursue their own lines of inquiry. Melissa indicated her opinion on this topic via a description of the primary classroom which she had just visited. While she was excited about the teacher's willingness to have Melissa teach science, she noticed that the theme for the year—fish—had been chosen by teachers, not children: "The school chose fish. . . I believe that's been pre-established. I think it's an organized thing. Although, the kids did not pick it."

Melissa's hope for involving children in curricular decision-making is given extra meaning when set up against Kevin's poignant story illustrating his understanding of the dangers of neglecting children's questions: "In college, we were presented with the option of doing research, or taking independent studies, but I had no idea what it meant. I had no questions that I wanted to answer, because science, all through my life, was memorizing, and just learning, learning, learning. Science was never an exploration for me."

Kevin values questioning in his definition of science and notes that his schooling did not support his own inquiry or his explorations of the world, but implicitly taught him not to question. He desires something different for his own students. Kevin is not alone in his demand that effective hands-on teaching be based in children's questions and interests and that children can even be responsible for the forms that their inquiries will take. Kathleen punctuates this point as she recounts her experience of inquiry in the elementary science education class: "It was good methodology to sort of, leave us to our own devices. By sort of giving us space as students who were able to devise our own methods, and the rules, thereby since we generated them, they're sort of more internalized."

Jeffrey has found that science talks (Gallas 1995) are a comfortable place for him to hand over choice to the children. He has struggled with his conviction that student choice in science is important, and he doesn't yet have the instructional repertoire to create settings in which he and the students can work productively together on mutually created questions. Science talks free him from the demands of teaching scientific knowledge, demands that he feels confident he could meet intellectually (because his science content background is strong) but not pedagogically. He introduced science talks into a summer science program for first through fourth graders in which he taught for several weeks: "[Science talk] is a structured way to hand over choice. Like, okay: Everyone want to talk about why the Earth's gonna blow up? I would have no idea, like what to do, like how to prepare a lesson for that! You know. But here's a structure. Like whatever they decide is a good question. Because all they're generating is a question, and that's what they need to contribute. So it's a task that they understand. *I* understand how to ask a question."

Jeffrey utilized science talks to practice choice-making for his students. I believe that he felt overwhelmed by the challenge of designing hands-on instruction that would attend to students' interests. The structure of the science talk allowed him to let go of control to a certain extent—he still chose the question from student-generated suggestions. He didn't need to plan complex activities or do a lot of research to support his students' learning in the *talking* processes of science. During science talks, his students were not actively engaged with physical materials or with nature. However, they were engaged intellectually, and this allowed Jeffrey the opportunity to practice the kind of choice-making that eventually may lead to more complex science explorations.

Some of these prospective teachers worry that when students are encouraged to engage in open-ended inquiry, necessary or interesting science content may be bypassed in favor of unstructured and aimless play. During one of our interview-conversations, Jeffrey and Barbara embarked on a very interesting discussion of when hands-on science became play. They decided, eventually, that they needed, as Barbara stated, "an interpretation of what playing is" before they could make much progress in this discussion. However, before they reached this point, Barbara argued against the assumption that science learning and play could not proceed simultaneously. Barbara defends her stance that engagement in science learning may look like play but is nonetheless indicative of children's involvement in the learning of the scientific content itself. She compared how the children behaved later in the unit, when they were pursuing work they had chosen with the materials, to before they had access to the hands-on materials, when their assignment was more restricted (naming electrical products): "The tables became their own little unit and they decided what went on. And they were *totally* engaged. I remember there was one time where I had them go and talk about all the electrical products. And I constantly remember telling the tables, like, 'Quiet voices. Did you figure out your products?' Whereas when this was occurring they would be like, 'Barbara, come and look at what we did! We got four batteries into one light bulb, look at how much brighter it is.'"

As I touched on above, Julia seems to have comfortably resolved the matter of student choice by considering the content of her science teaching to be scientific processes. She describes her science teaching, in response to a question from Kevin about finding appropriate activities:

"The way I did the activities was I came in and first with the chemistry just brought a lot of supplies, and I had copied experiments out of books for kids. And I let them choose whatever experiment they wanted to do. And in class, we just did the experiments. And if they wanted to alter those a little bit, and experiment and play around, try mixing different things, they could. And they compared, and they just got excited by it. And then, after they had sort of gone in, jumped into it, then I started introducing things like documenting what you'd done. How to make observations and chart what you've done."

Because Julia is concentrating on teaching science processes, she does not worry about her own lack of content knowledge in the field of chemistry (although she stated that she was learning more in this area). When giving children choice, it is certainly important that the teacher have a strong enough background to introduce powerful scientific ideas to further their learning. It seems that, if we are serious about basing curriculum in children's questions, it will be necessary to at least occasionally take the risk of being inexpert and, indeed, learning along with the children.

Unpacking Hands-on Science: Summary

In summary, I have realized that the attractiveness of hands-on science teaching involves much more than merely keeping the children busy and avoiding confrontation with one's felt ignorance. The choice is both more reasoned and more complex than that. These preservice students are choosing hands-on science teaching because they want children to have quality time with the physical world in a form that they can get to know personally; the distances implied and actualized via textbooks and lectures are traded for direct engagement with the world. The rigidity and isolation, as well, of the detached mode of instruction is discarded. Children in these students' ideal classrooms would be creating and addressing their own questions, moving among playing, making, talking, drawing, and other form of human expression that would aid their explorations.

I believe that these preservice students' descriptions of hands-on science teaching fit squarely within a representation of science as everywhere. Hands-on science teaching, in many of their eyes, incorporates much more than laboratory work with manipulable materials. In addition to signaling an escape from text- and lecture-based science teaching, this type of *physical* science becomes something larger: It begins to signify interactions between children and nature, ones that involve the whole body, the whole psyche, and the whole of the social complex within which we all live and learn. When taken to its logical conclusion, the motto "science is everywhere" causes one to insist that the personal, the social, and the natural are inextricable. The sloppiness of this concept may be its strength.

Conclusion

In my own teacher preparation program, I was a student in a course for people becoming certified to teach in secondary settings. The professor had engaged us in a discussion of the text of the daily horoscope. I can't remember why he did this; what I do remember is that the concept of fate had entered the discussion. Still high on my own self-assured spot in the hierarchy as a scientist, I said to those students in the back row who were so involved in this discussion, "You people obviously aren't scientists." Recalling this obnoxious declamation is mortifying; I am only doing so now because I know that not so many years ago I would have thought the same of my own students, and the elementary teachers they will become. Luckily, largely through feminist teaching and feminist teacher research, I have learned the importance of looking into my own assumptions and prejudices. Working with preservice elementary teachers has supported and pushed my effort to see beyond my own expectations.

Local as well as national standards and tests urge elementary teachers to teach more science in their classrooms (AAAS 1993; AAAS 1990; NRC 1996). These top-down efforts, accompanied by resurgence in the use of standardized testing, are not going to go away. On one hand, insisting that elementary teachers represent science in prescribed ways reifies the stereotypes that they know quite well enough from living in a society that depicts science as elitist, its knowledge as arcane and inaccessible, and its practitioners as super- or nonhuman. On the other hand, hands-on science is not a likely candidate for toppling the hierarchy. However, in my view, these students' conceptions mesh well with the contemporary standards movement toward authentic, original scientific inquiry as the basis for classroom instruction (AAAS 1990, 1993; NRC 1996).

The majority of preservice elementary teachers tell me that they believe that "anyone can do science," and they hope to create classrooms in which this principle prevails. This apparently prosaic statement becomes notable when one remembers that our schools and the Western culture and heritage they represent transmit an opposing message, and that the still-predominant images of scientists and science do not include everyone. Nonetheless, we can refuse to encourage prospective teachers to buy into the "Ain't science grand?" invention that blames outsiders for their failure to profit from its actual and mythologized grandeur. At the heart of my teaching and research efforts is a desire to open up science and science education—to suggest alternative faces and approaches to studying the natural world—approaches that have been hidden behind the construction of science as a purely rational, asocial, and dispassionate enterprise. It is within this context that I tend to view the elementary science education course, my research, and my students' thinking about science education. I have been frustrated in the past by expecting that students would adopt *my* view of the journey they should take through autobiographical reflection and critical analysis, coming out on the other side with a clear-cut and passionate view of what it means to teach "science for all" in an unjust society—in other words, not only adopt my view but adopt my commitments! Recently, I have found it much more rewarding to set aside a concern for my success in helping students attain these specific goals. As I study their visions of science teaching, I can better see the intelligence, optimism, and originality that they bring to this enterprise. While the great majority of the students in this course have not yet developed an understanding of the role that school plays in maintaining oppression and punishing difference, they want science to be communal, nurturing, embodied, and connected for their students. They hope to create classrooms in which these ethics prevail. They want their students to learn scientific content, but they want it to happen in an atmosphere that realizes a kind of scientific activity that welcomes the whole self, welcomes questioning and open exploration, and denies the inaccessibility of scientific knowledge and practice to all but a few of the social and intellectual elite.

References

American Association for the Advancement of Science (AAAS). 1990. *Science for all Americans*. New York: Oxford University Press.

AAAS 1993. *Benchmarks for science literacy*. New York: Oxford University Press.

Barton, A. C. 1998. Teaching science with homeless children: Pedagogy, representation, and identity. *Journal of Research in Science Teaching* 35 (4): 379-94.

Bell, B. 1995. Interviewing: A technique for assessing science knowledge. In *Learning science in the schools*, ed. S. Glynn and R. Duit, Lawrence Erlbaum Associates. 347-64. Mahwah, NJ: Lawrence Earlbaum Associates.

Bleier, R. 1986. *Feminist approaches to science*. New York: Pergamon Press.

Bloom, J. W. 1998. *Creating a classroom community of young scientists: A desktop companion*. Toronto: Irwin Publishing.

Driver, R. 1989. The construction of scientific knowledge in school classrooms. In ed. R. Millar, Doing science: Images of science in science education, 83-106. New York: Falmer Press.

Fausto-Sterling, A. 1992. *Myths of gender: Biological theories about women and men*. New York: Basic Books.

Frye, M. 1992. The possibility of feminist theory. In *Willful virgin: Essays in feminism* 59-75. Freedom, CA: Crossing Press

Gallas, K. 1995. *Talking their way into science: Hearing children's questions and theories, responding with curricula*. New York: Teachers College Press.

Gardner, H. 1983. *Frames of mind: The theory of multiple intelligences*. New York : Basic Books.

Griffin, S. 1978. *Woman and nature: The roaring inside her.* New York: Harper & Row, Publishers.

Grumet, M. R. 1988. Pedagogy for patriarchy: The feminization of teaching. In *Bitter milk: Women and teaching*, 31–58. Amherst: University of Massachusetts Press.

Hartsock, N. 1979. Feminist theory and the development of revolutionary strategy. In *Capitalist patriarchy and the case for socialist feminism*, ed. Z. K. Eisenstein, 56–77. New York: Monthly Review Press.

Hazelwood, C. C. 1996. Shaping identities in school science: A narrative study of girls of Mexican origin. Ph.D. diss., Michigan State University.

Howes, E. 1998. Connecting girls and science: A feminist teacher-researcher study of a high-school prenatal testing unit. *Journal of Research in Science Teaching* 35 (8): 877–96.

Hubbard, R. (1994). *Profitable promises: Essays on women, science, and health.* Monroe, Maine:Common Courage Press.

Keller, E. F. 1983. *A feeling for the organism: The life and work of Barbara McClintock.* New York: W. H. Freeman and Company.

———— 1985. *Reflections on gender and science.* New Haven, CT: Yale University Press.

Krajcik, J., Berger, C., and Czerniak, C. 1999. *Teaching children science: A project-based approach.* Boston: McGraw-Hill College.

Meier, D. 1995. *The power of their ideas: Lessons for America from a small school in Harlem.* Boston: Beacon Press.

Merchant, C. 1980. *The death of nature: Women, ecology, and the scientific revolution.* San Francisco: Harper & Row.

National Research Council (NRC). 1996. *National science education standards.* Washington, DC: National Academy Press.

Pomeroy, D. 1994. The lesson of the country bunny, or breaking Western scientific tradition. Manuscript. Beaver College, Philadelphia, PA.

Roth, K. J. 1989. Science education: It's not enough to "do" or "relate." *American Educator* 13 (4): 16, 18–22, 46–48.

Roth, K. J. 1995. Stories of alienation and connection: Examining the neighborhood of science from the margins. Manuscript. Michigan State University, East Lansing, MI.

Stanley, L., and S. Wise. 1993. *Breaking out again: Feminist ontology and epistemology.* New York: Routledge.

Tutt, R., and B. Rosenthal. 1995. Legos-based mathematics teaching and learning in a middle school: First gleanings. Paper presented at the First Annual PDS Research Conference, March, East Lansing, MI.

Walkerdine, V. 1992. Progressive pedagogy and political struggle. In *Feminisms and critical pedagogy,* ed. C. Luke and J. Gore, 15–24. New York: Routledge.

Notes

1. In this research, students were given the option of my using a pseudonym or not. All of the names I've used in this paper, with the exception of Kathleen, are pseudonyms.

2. The majority of these quotes were compiled with my adviser and teaching mentor in graduate school, Kathleen J. Roth. I have added a few quotes, but this activity originated with her.

PART TWO
SECONDARY SCIENCE EDUCATION AND SCIENCE TEACHER EDUCATION

Chapter 7

Re/Writing Science from the Margins

Gaell M. Hildebrand

> Discourses on appropriate genres in science
> Avoid questions of whether the conventions
> Inculcate a positivist view: an over-reliance,
> And ought to be the subject of contravention.
>
> There is no "scientific genre"—as a singularity,
> And to insist that school writing be in compliance
> With restrictive styles is to effectively guarantee
> Limitations on learning through writing in science.

Discourse and Power

> Discourses are about what can be said, and thought, but also about who can speak, when, where and with what authority.
> —Stephen Ball,[1] *What is Policy?*

By using the concept of a discourse you can clearly see that in science teaching there are shared values and assumptions about whose voice is heard, what is taken for granted, what activities "naturally" lead to learning science, and which writing genres are deemed appropriate for displaying learning. But discourses are not benign patterns of language use; they are contestable areas, where power and hegemony are pervasive or, in competing discourses, disruptive. In this chapter I show you that the discourses that re/present[2] science through writing are built upon foundations that can be legitimately challenged from several critical standpoints.

I have deliberately broken the conventions of academic discourse in the framing of this chapter. By presenting the ways that some teachers are re/writing science I have chosen to allow form to metaphorically indicate substance at various points. For example, I have written poetic interludes where it seems appropriate for me to do so. I do this as a deliberate project to disrupt the established practice of separating the discourses used in the arts from those used in the sciences (Charles Snow 1963).

My purposes in using poetry are to evoke feelings, prod reflections, act politically, enjoy complexity, use metaphorical thinking, shift gear, and make connections. I use poetry despite being warned: "use the linguistic and stylistic resources of the poet or artist for scientific communication, and you will not have standing as being scientific" (Jay Lemke 1995, 178). I am prepared to take the risk. I choose to play with the linguistic conventions in order to illustrate, by form as well as content, my point that our pedagogical practices in science teaching are informed by our belief systems, our frame of reference, our past experiences, our aesthetic appreciation: our personal subjectivities.

I have also formatted two short sections of this chapter as competing discourses: where the voices presented in the left margin, both literally and metaphorically, offer an alternative view to that generated through hegemonic "scientific" discursive text, exemplified by the right column. The shifting and multiple points of critique of the dominant mode of writing science are indicative of arguments for a re/writing of science. I have selected this series of snapshots as a representative collage showing reasons for, and consequences of, a re/writing of science, particularly as it is used with/in science classrooms. My hope is that new spaces may be opened up for critique of conventional writing practices in school science—spaces that may highlight not only that those marginalized by science can be mainstreamed when a re/view of science writing is facilitated, but also that the process can simultaneously change the mainstream of learning science itself.

* * * * *

My perspectives on writing, learning, pedagogy, and science have grown out of who I am. I think it is therefore important to write myself into this story of why I think that disrupting the writing practices used in learning can be one means of facilitating the entry into science of those who have become disenfranchised from it.

> Girl.
> A privileged childhood,
> Living on a farm with
> Half a dozen sisters and a brother;
> Almost always having fun, all of us,

Working for the family's good,
Milking cows, chopping wood.
Crossing borders
Meant driving over the river and going interstate.
Changing power
Meant grid electricity arriving at our farm gate.

Teenager.
A confused youth:
An A student with aspirations of an academic gown,
Unknown in my family, rare in my town.
Only three girls in my final year physics class
Feeling numb in that chilly atmosphere,
Falling in love made it harder to persevere.
Crossing borders
Meant choosing to specialize in physical science.
Changing power
Meant independence and self-reliance.

Young white woman.
With optimism I moved to the city
And took my science degree,
With two majors:
One in physics, one in chemistry.
I also studied science history and philosophy,
At last: a space to challenge dogma, rules and theory.
Crossing borders
Meant friends from Vietnam suffering dislocation.
Changing power
Meant Germaine Greer and women's liberation.

Beginning teacher.
Girls telling me "science is boring."
But by creatively exploring
Science through sharing, acting
Talking, writing and drawing,
We overcame some frustration,
By trying new means of communication.
Crossing borders
Meant using teaching strategies borrowed from the arts.
Changing power
Meant listening to students' words, feelings and hearts.

Experienced teacher.
Relief, joy and excitement:
The McClintock Collective forms!
Women working together to re/form,
Struggling towards gender inclusive science,

Visioning ways to move beyond positivism
Re/constructing power and gender, discarding dualisms.
Crossing borders
Meant transitioning into teacher professional development.
Changing power
Meant statewide curriculum change a shared accomplishment.

Teacher educator.
And thus was this interruptive chapter conceived.
No distanced researcher here, but one who perceived
The potency in pursuing your passions,
Remaining whole, not broken into fractions.
As a researcher (like others) I act on my values and desires:
Seeking pedagogies that challenge inequities, that inspire.
Crossing borders
Means strength in holistic perspectives, multiplicity, desires.
Changing power
Means problematizing science, teaching, gender: lighting metaphorical fires.

Postscript:
    This chapter would be entirely poetic,
    But my long engagement in things scientific
    Makes it difficult to challenge the genre condoned
    When my expository writing skills are more finely honed.

* * * * *

The Politics of Disruption

I see the world from a feminist frame of reference. When considering what I call the politics of disruption, it becomes clear that those who are in some way marginalized by established discourses and practices have a greater personal investment in fractures or interruptions: anyone constructed as the "other."

> Marginalized as the other:
> Not white,
> Not male,
> Not heterosexual,
> Not speaking English as a first language,
> Not living in privileged neighborhoods.
> The other
> Has become the majority.

* * * * *

The Other also includes most of the world's teachers!

But not all people who have become marginalized on the verandahs of science recognize that they have been restricted from having access to power. The way that discourses of power operate is to "domesticate" people into believing the dominant view, even when they are otherized themselves. Joan Wink argues that this support of hegemony occurs because "those who are marginalized by the hidden processes" either "haven't taken time to reflect critically on themselves and their roles" or they see playing by the dominant rules "cynically, as the superhighway that must be used to get ahead" (Wink 1997, 43).

I speak here from the perspective of one who explores the education-gender system (Jane Roland Martin, 1996) as a source of disenfranchizement from science and from power. Some writers who challenge hegemonic science pedagogy speak from a perspective of disadvantage due to multiculturalism-education (e.g., Lynne Hewlett 1996; Peter Esterhuysen 1996; William Stanley and Nancy Brickhouse 1994; Mary Atwater 1994, 1996), while Paulo Freire (1968, 1985), Adriana Hernández (1997), and others who speak with anti-Colonialist voices, speak from perspectives of political and economic oppression. Each perspective has an investment in a disruption and a re/writing of the dominant discourse. I, like many others, have an ideological investment in providing a competing "intelligible, plausible and fruitful" discourse (Peter Hewson 1996, 133; also 1981) as I seek a conceptual change in science educators' frameworks about the interactive science/pedagogy nexus.

At times there appears to be an impenetrable barrier to change, built from the ideological position that current practice is the natural order. But interrupting and contesting dominant discourses, as history shows us, is always possible—even in science. "We must all, women and men, become aware of the power dynamics which are discouraging for women and, most importantly, work to change that atmosphere"; this implies that "our role as teacher educators and science educators requires that we inform ourselves on these issues, and actively pursue change" (Sharon Haggerty 1995, 7). As Haggerty argues, it is the responsibility of each of us to change the inequities of the current situation; we can no longer leave it to the marginalized groups, the victims of otherizing, to right the wrongs of hegemony. All science educators (researchers and practitioners alike) must become informed and act for change if we are not to be complicit in maintaining power for some and marginalization for most.

Changing power on a widespread scale will require large numbers of teachers to take many small steps toward more liberatory pedagogies. Managing to interrupt and disrupt current practices in ways that are manageable requires conceptualization of a new pedagogy, one that allows teachers to *think ideologically while acting practically* (see Hildebrand 1998) so that they can change restrictive writing and learning practices.

> Our discourse on pedagogy
> Is shaped by our frame of reference,
> Our ideological positioning.
> Through my feminist lens,
> I re/vision ways of learning science:
> By interrupting the hegemony
> Of science-as-usual,
> Of power-as-usual,
> Of pedagogy-as-usual,
> Of writing-as-usual.
> Changing power
> By
> Crossing borders.

* * * * *

Available and Allowable Genres

In schools, science is re/presented through the discourse patterns of the texts that are *available* for students to read and *allowable* for them to write. I argue that the writing genres available for students, as consumers of science knowledge, construct science as positivist: a rational field of knowledge that inductively "discovers" through the so-called scientific method, a discourse that can be traced back to Frances Bacon (1561–1626) and that was reinforced by Auguste Comte (1798–1857). The available texts imply that there is a right way to write science: implicitly the texts model scientific writing, and explicitly they dictate writing models for the students to use, models that follow an inductive, empiricist view of science.

Another contestation I have with the available texts in school science is that they construct science as masculine. This has been amply demonstrated as a worldwide phenomenon by many papers presented at the international Gender and Science and Technology (GASAT) conferences over the last two decades. Gender analyses of science textbooks, for various levels and subjects, show that science is portrayed as masculine through pictorial images; particular scientists mentioned; analogies and metaphors used; applications drawing on common male experiences; and the language used, such as the scientist as "he" (e.g., Judith Bazler and Doris Simonis 1989). Parallel analyses based on race or language background would, I predict, construct an image of science as an area of human endeavor only undertaken by white English-speaking males.

The writing genres usually *allowable* for students to produce in school science form, or construct, an image of what counts as science learning. When this writing consists entirely of non-narrative genres (note taking/making, summaries, glossaries, reviews, calculations and problems, worksheets, and laboratory reports) then science is perceived as factual. On the matter of writing up the ubiquitous report, Clive Sutton says that "practising how to write a scientific report . . . is a reasonable part of science education, but not every day!" (1992, 89).

Using Hybrid Genres: Imaginative Writing

> Science does not suppress the imagination; much of it, like the arts, springs from the imagination.
> —Hanbury Brown, *The Wisdom of Science.*

I use the term "imaginative writing" to include a spectrum of writing styles that draw on the imagination. This type of writing is most frequently called creative writing by science teachers in Victoria, Australia; but local English teachers associate that term with fiction/narrative and so I have chosen to use a new term. Under the broad term of imaginative writing, I incorporate any form of writing that involves a blend of the creative or the expressive with non-narrative genres. I have referred in the past to this writing in science as "non-conventional writing" (Hildebrand 1991), but I realize that it is not a term that will ever take off with teachers! Perhaps the most technically correct term would be "hybrid creative-scientific writing" because it would signify that the writing is an unusual blend of these written forms.

There are advantages to not defining too precisely what I mean by imaginative writing in the context of secondary school science pedagogies. As Christina Hart points out, "from a reformer's point of view a certain amount of vagueness is a strength" (1995, 347) because if you define something too narrowly, many people will feel immediately excluded from the notion and turn away from a site of potential interruption and change. "By allowing a penumbra of ambiguity around the boundary of the innovation, interest groups of many kinds can see within it ways in which it might serve their different needs" (David Layton 1994, 32).

For the purposes of this chapter I define imaginative writing as any hybrid, or blended, genres that use scientific and/or non-narrative genres (i.e., recounts, procedures, reports, explanations, expositions, discussions, etc.) in conjunction with other fictional/narrative genres. This provides for a "penumbra of ambiguity" that may facilitate teachers' development of alternative writing tasks beyond those described here.

In our postmodern world the existence of facts is very much contested: there are alternative beliefs about the world, no fixed certainties, no definitive truths, no facts. Patricia Rowell expresses her, and my, concern that "non-narrative writing is referred to as 'factual' writing" by many linguists and teachers. She points out that "the everyday usage of the word 'factual' as objective and uncontested information" is likely to contribute in schools to an image that such genres of writing are about the truth (1996, 30). "The portrayal of science as an accumulation of undisputed 'factual' information is a setback for those who would characterize science as a human, interpretive activity" (1996, 32).

The allowable genres for writing to learn science can be considerably expanded, as can the conception of science itself, under the rubric of imaginative writing. The following examples serve to open a window of possibilities:

- Imagine you are a water molecule: describe five changes of state that you have recently gone through (anthropomorphic narrative).
- Write a conversation between two electrodes in a Galvanic cell, describing what they see happening around them (conversational narrative with anthropomorphism).
- Prepare a travel brochure advertising a trip to another planet (advertisement).
- You are a parachutist (dancer, basketballer, etc.); describe the forces acting on you throughout a jump (dance, game, etc.) (recount).
- Write a poem describing a monotreme; write it in the shape of that animal (poetry).
- Write an ode to the halogens (poetry).
- Write an obituary for the last living tree (journalistic).

Generic writing products that fall under this rubric include the following:

- a wide range of narratives/stories (e.g., conversation, fantasy, futuristic/speculative fiction, mystery, adventure, romance, cartoon strip, detective story, and anthropomorphic narrative);
- journals (e.g., autobiography, diary, travel journal);
- letters (e.g., postcard home, letter to the editor, Dear Dorothy Dix letter, etc.);

- poems (e.g., ballad, ode, limerick, cinquain);
- lyrics (e.g., rap, rock, ballad);
- scripts (e.g., screenplay for a live production—may be video taped—and for a radio show—may be audiotaped);
- advertisements (e.g., travel brochure, jingle, poster);
- and journalistic writing (e.g., news, sports report, editorial, feature article, obituary).

Further specific examples of imaginative writing are described by Leonie Gianello (1988); Sue Lewis and Anne Davies (1988); McClintock Collective (1989); Liz James (1989); Gaell Hildebrand (1989); Sue Stocklmayer (1989); Vaughan Prain (1995); and Sue Avery and her colleagues (1996). Table 7.1 lists the writing tasks that some of the teachers who collaborated with me in a research project used in their secondary science classrooms.[3]

Table 7.1 Writing Tasks Used by Collaborating Teachers

Sample Writing Tasks	Grade	Science area	Type of writing product
Michael			
Choice of three tasks:			
Imagine you are a journalist and can travel in time: Interview a chemist and write an article for *The Globe*.	9	chemistry	newspaper article
You're invited inside an atom.			story (anthropomorphic)
Imagine you are a water molecule.			
It's "bean" a long time. Give advice to a broad bean seed on its growth needs.	7	biology	poetry or story (anthropomorphic)
Write a cinquain or shaped poem about a water droplet.	7	chemistry	poetry
"Honey, I shrunk the kid." Traveling through the ear.	9	biology	story (anthropomorphic)
Nicole			
Write a story where a crime is solved using forensic science techniques.	9	integrated unit	detective story
Write a story about your travels as a recyclable object.	8	environmental science	story (anthropomorphic)
Write a story about your life as part of a plant.	7	biology	story (anthropomorphic)

Table 7.1 Continued

Sample Writing Tasks	Grade	Science area	Type of writing product
Kit			
The water story (changes of state).	7	chemistry	story (anthropomorphic)
Design and describe a space station, showing how you could live on a particular planet.	7	astronomy	story
Mystery powders.	7	chemistry	detective story
"Hello, I'm Cindy the Cell."	7	biology	story or cartoon (anthropomorphic)
Alex			
You are Roula, the red blood cell. Describe a day in your life.	9	biology	story (anthropomorphic)
Yesterday was a freaky day; I woke up to find myself in a world without friction.	9	physics	story (fantasy)
The circulatory system as a city transport map.	9	biology	annotated map with metaphors
Describe a trial to determine if something is living or not with arguments for the defense and prosecutor.	7	biology	legal report
Katie			
Six choices, including Prepare a travel brochure for a part of the solar system. Write a letter home from space.	7	astronomy	advertisement letter
Write a song about four elements.	7	chemistry	poetry/song
Write a book for primary students on safety with electricity.	8	physics	manual
Sandy			
The fate of a ham sandwich.	9	biology	story (anthropomorphic)
You are a molecule in crude oil; describe your progress through a fractionating column.	10	chemistry	story (anthropomorphic)

Teachers' Voices on Hybrid Genres

From a poststructural feminist standpoint, I argue that our thinking is shaped by the discursive patterns within which we feel confined to speak and/or write. By diversifying writing, we are freeing our students from the restrictive modes of thinking that science has imposed on its learners in the past. Brian Hand and Vaughan Prain argue that by using a range of writing practices, teachers can ensure that writing becomes an "active process of synthesizing and reprocessing knowledge," and this "extends the students" beyond the mere reproduction of knowledge common in conventional writing tasks (1996, 26).

Anthropomorphic narratives

When writing anthropomorphically, students adopt a human-type persona even though they are imagining they are a nonhuman object. Many anthropomorphic narratives follow this pattern: "You are a [fill in the blank]; describe what happens to you when you [blank]." Another form, frequently used by my collaborating teachers was: "You are a [blank]. Write a letter [or series of postcards] home from [blank] describing your journey through [blank]."

> The 'Postcards from a trip down the gut' gave me a way of knowing that they could picture themselves inside the gut, and from the food's point of view see the villae and know about the secretions and know whether they would have to be in a special suit to avoid acid in certain places, and things like that. (Yvette)

> You have to have a pretty solid understanding of the terms you're using, and a good understanding of the scientific principles you're talking about, so that you can write in an anthropomorphic way. (Rosemary)

This notion of the power of anthropomorphic narratives to allow students to get inside the science concepts is a repeated message from these collaborating teachers.

> I think it helps students learn because the key is that they're identifying with a concept to use it as part of their story . . . If you have to become part of something then you're developing your understanding fairly closely. (Nicole)

Mike Watts and Di Bentley give examples from different science classes where the teachers speak anthropomorphically in order to help their students understand concepts, for example, "the particles in a heated solid vibrate more and more until they cannot stand it any longer and they melt" (1994, 88). Much of the metaphoric language used in science is anthropomorphic in origin, for example, "hydrophobic" molecules and "electrophiles." Clive Sutton reminds us that teachers should feel comfortable using anthropomorphism as "there is surely no need for us to be apologetic about forms of expression which resemble those used by scientists in their creative effort" (1996, 5). Watts and Bentley argue that an increased use of anthropomorphic writing "would increase the 'human' appeal of science and reduce the alienation felt by young women in science," and they urge teachers to "exploit [the] pedagogic potence" of anthropomorphism to help sustain girls in science (1994, 83).

Journalistic writing

A variety of journalistic forms, from news reports to obituaries, have been used by some of the teachers in my study. The teachers frequently using journalistic genres as a mechanism for linking students' social worlds with their science learning.

> They really love talking about things that are of immediate impact to them, things that affect their daily lives, and everybody knew somebody who had some [medical] condition, or they had it themselves and that seemed to interest them. And I thought, "well, OK, why don't I get them to explore something that interests them?" They then had to interview someone about their condition. I thought it was a good way of bringing in technology as well: How's the condition managed? What tests are done to diagnose the condition? How does it affect the family? The lifestyle? They had to bring in a whole lot of issues like that into their writing. (Katie)

Only in Katie's case did I hear a prewriting discussion about using some specific strategies of a journalist to capture an audience and hence to deconstruct the ways journalistic writing can be used to persuade: "I also get them to do a media analysis. We analyse an article from New Idea [a women's magazine] and from The Age [a newspaper] and it has to do with someone's condition. . . . We talk about the tools that the writer uses to engage an audience and how the message is being communicated. . . . and it makes them much more aware of how journalists operate."

Scripts

Some teachers associated writing with another activity in the topic. For example, some used role play and dramatic representations to help students learn science concepts, and the scripts that were written for performance are another form of imaginative writing. The performances may simply be for their own class or they may be as extensive as the production level required for the annual "Science Drama Awards" run by the Science Teachers' Association of Victoria (STAV).

> I also involve drama in there because when we do the improvisations with students I also get them to write them down in a dramatic form. So they do script writing for me in that way. (Anastasia)

Poetry and Songs

This is a cinquain poem of mine:

> Science,
> Powerful, creative, evolving,
> Challenging, exploring, thinking, learning,
> Writers actively constructing meaning,
> Science.

The succinctness is appealing to several teachers who mentioned it as a useful vocabulary-building tool that is also a nonthreatening task for their students.

> I love those little poems—the cinquain poems. Because they really seem to knock it on the head: it might be the concept of change or flight, or something. It allows the kids the chance to creatively think about the words that describe that concept without having to go into a long explanation. . . . I think it's very freeing to have an opportunity to discuss and explore what you understand by a science concept through just picking out some other words and writing a short poem. (Jeff)

Katie encourages her students to write songs (poems set to music): "They do a song in year 7, working in groups, and they've each got four elements. So they've got to do research first. . . . Then I set them the task and I say to them, "OK, you're going to be making up a song from this information, that you're going to perform." . . . We discuss, "what might people want to know about these elements?" We write up a list of questions and then they go to it and they find the answers [and] eventually they do produce something reasonable together. And I've got some videos there of them actually performing it. . . . and they don't particularly recognize that they're writing verses, or poetry."

In this case, as in many others, the written product created provides the students with a purpose for their learning and acts as a motivational device.

Fantasy—touched by realism

Many of the tasks that teachers set are of the type used by Yvette when she asked her students to write about a frictionless world: "They had to write what that would mean to them in terms of explaining four situations where they'd normally rely upon friction to actually do a movement. So they had to research what the thing was that they would do and how friction would help them normally function, such as walking across the floor in their bare feet, and then comparing that to a situation that they could imagine where there was no friction . . . only by them showing me what would happen if friction wasn't there did I know that they knew what friction was." As Yvette explains, sometimes the absence of a scientific concept can only be possible in a fantasy world, and yet its absence can become a stimulus for thinking more deeply about its presence.

Some Teachers' Reasons

Writing with imagination
May lead to an early cessation
Of exclusions and frustrations
Through a contestation
Of the dislocation and alienation
Of several generations
Arising from the mystification of
Science.

* * * * *

All but one of the collaborating teachers in my project began using imaginative writing in their science classes because of an active interest in the notion of an inclusive curriculum. In such a model, teachers' efforts are focused on delivering their teaching in such a way as to include all students and not alienate sectors of their classes. In particular, the three groups who were deliberately targeted—at least initially—by these teachers were:

- girls;
- students whose heritage, or family, language is not English;
- students whose academic levels were not high and who were generally perceived as alienated from schooling in general, and from science in particular.

Teachers' Reasons for Using Imaginative Writing

(Voices in the Left Margin)	**Conventional Science Writing**
	. . .
"We started because we were working with Rowena, and the Special Ed people. We were focusing on language and that started us off. It began more as an ESL [English	as the object moves on the frictionless

as a Second Language] thing, I think, and then from that we looked at it from the girls' perspective." (Maria)

"I have a very challenging context for this, because I was working with [a named suburb] kids, teaching science to students who came from a variety of dislocated backgrounds [and] who had all sorts of social and economic constraints . . . that had impeded their educational history. So for some of them writing anything was a challenge let alone the traditional approaches to writing up a science experiment. So I was really challenged in a major way, and I think that had a powerful influence on me . . . And obviously writing was really powerful and the diversity of it [pause] there was a whole range of ways of opening up writing styles." (Elaine)

"Well, I think actually, I first came across it in my very early years of teaching when I read an article published in *Readers Digest*, and it was relating to body systems and it was done in a really imaginative way . . . And I just thought 'what a good idea'. [But] when I first started teaching, the convention was to write formal 'prac.' reports and science was a pretty formal subject. I think if you did something like that, and I'm looking at 20 to 30 years ago, you would've been laughed out of the place, and that wouldn't have been regarded as science, because it just wasn't done . . . The first time I used it properly was when I attended a McClintock[4] in-service and since then I've used it consistently." (Michael)

"I was just thinking of that poster with Einstein that's up in the chem laboratory: that 'imagination is more important than knowledge.' I guess what we'd encourage kids to do would be to start thinking, and not stop thinking, about stuff that they come across in the lab. And not to think about science as a collection of 'facts' that haven't changed and won't change. And to think that it's more of a dynamic process—and I think by doing things that encourage kids to write and think and talk divergently, by using kids talking to each other about their work, by using creative writing, by using role play, we're encouraging them to move away from the stereotyped image of what science has been, and to think about it in a more evolutionary way." (Alex)

"I think that probably most good scientific discoveries, most particularly important ones, have been because someone has crossed a barrier—because something has appeared to be what it ought not to be, or because somebody different has looked at a result with different eyes, or somebody has shown a biological specimen to a chemist, or whatever." (Yvette) . . .

surface it
remains at a . . .

. . .

photosynthesis
involves light
acting as a
catalyst so that

. . .

while the red
blood cell carries
the oxygen around
the hemoglobin
acts as . . .

. . .

oxidation occurs
at the anode in
the cell because

. . .

the common
structural
characteristics
mean that . . .

"My original premise when I started teaching science was that 'science is for all', it isn't just for people who were going to be studying at university or whatever. I've always thought that science is empowering for people . . . That's the whole point of this: I'm trying to make things more accessible." (Katie)

"We learn through connecting to previous experience, and I think science has really disconnected from all of that, and so it really built, unwittingly, on a masculine view of the world. I think we've been breaking that down, and opening that up, and now it comes out in the way that we teach . . . [and] the history of science has been much more extraction and disconnection and the learning of abstract knowledge and ideas. So I think it's trying to challenge and extend and diversify the meaning of science." (Elaine)

"It sounds ridiculous but it was almost a breakthrough in my teaching. Seeing that some people need to learn in that way, but not everyone does, and seeing that you had to give that opportunity to some kids to be able to do it, to be able to learn it. There was no point in me standing up at the board and talking about this sort of stuff to some kids. That was never ever going to be a way I would help them get some ideas into their heads." (Sandy)

"I think this is really where it shows up that we've failed in the old way that science was taught; it's the most clear indication that it didn't work. Because we cannot interpret what things mean for someone when they ask us. 'You're clever, you did science, you should understand this.' How many times have you been in that situation? And you haven't had a clue, because your knowledge is so compartmentalized, and you also aren't good at verbalizing what you know? If you haven't practiced it in the science classroom and you've just stuffed in all this knowledge, in little bits, it's not going to squish around and come out as a nice cohesive message years later." (Chris)

. . .
the aperture
generates
diffraction
patterns that can
. . .

. . .
the third sample
was rendered
unusable and it
was discarded so
that it . . .

. . .
conventional
current
*flows towards it,
while actual
current flows
away from the*
. . .

. . .
determined by
observing the
rate of decrease
of the absorption
. . .

Writing in/forms Science

David Locke (1992) sees "science as writing" because the key tangible artifacts of science are the written traces left behind by scientists. The interactions between science and writing are two-directional. Writing is *both* the record of science practice, *and* writing shapes the discourse of the community by constructing the paradigms of Thomas Kuhn's (1970) "normal" science.

> Writing in/forms science.
>
> Writing in science
> Provides the artifact,
> The trace left behind.
> Science as writing.
>
> Writing forms science,
> Constructs its shape,
> Giving it direction.
> Science embodies writing.
>
> Writing informs science.
> Illuminates thinking,
> Generates new insights.
> Science uses writing.
>
> Writing in/forms science,
> Recording and constructing,
> Defining the paradigms.
> Science valorizes writing.
>
> Analogously
> Writing in/forms science learning.

* * * * *

Writing in Science

In the same year that Charles Snow (1963) wrote his *Two Cultures* treatise, claiming distinctive features for the arts and the sciences, the Nobel laureate Peter Medawar (1963) wrote his classic exposé of the condoned writing practices in scientific papers. He recognized that papers construct a fraudulent view of science, clearly dissonant with the processes of science and its knowledge production projects. Medawar said, "hypotheses arise by guesswork" and they are most certainly "not of logic" (reprinted in Bernard Dixon 1989, 173). Medawar also argued that "the scientific paper is a fraud in the sense that it does give a totally misleading narrative of the process of thought that goes into the making of scientific discoveries. The inductive format of the scientific paper should be discarded" (quoted in Dixon 1989, 175).

David Locke argues that while many scientists genuinely believe that "their language does not much matter, that it is merely the empty vessel into which the content of their scientific thought is poured," this is not the case (1992, ix). Locke argues that previous discourse analyses of the writings of the poet/author and the scientist have concentrated on the differences, trying to "otherize" the writing styles, seeking a legitimation of one form as superior or more powerful. He meticulously deconstructs the oppositions that supposedly underpin each form, showing that their similarities are greater than their differences. For example, literature and scientific writing both use similar persuasive tactics in an attempt to convince the reader of a particular point of view. Locke also illustrates the expressive and imaginative side of scientific writing, its artfulness and affectivity, and exposes the way it indicates the social milieu of the writer(s). Locke challenges the notion that scientific language should retain its "privileged position as pure functional notation, as mere shorthand records" because of the literary devices that clearly exist in the writing (1992, viii).

In Locke's analysis of the original letter to *Nature* on the structure of DNA (Jim Watson and Frances Crick 1953), he showed how the techniques of literature were well used. For example, understatement was used to imply, but not state, claims: a clever use of this literary tool that highlights a point. Crick later discussed how their first nine-hundred-word letter was "brief and restrained" but that he did talk Watson into adding the famous understatement: "It has not escaped our notice that the specific pairing we have postulated immediately suggests a possible copying mechanism for the genetic material" (from the *Nature* paper, cited in Crick 1988, 66). Crick said this sentence was a "compromise" between his own desire to make a "claim to priority," being the first to notice and not wanting others to think they had been "too blind to see" the possibility, and Watson's "periodic fears that the structure might be wrong and that he had made an ass of himself" (Crick 1988, 66). This concurs with Bruno Latour's observation that when authors are "on dangerous ground, understatement proliferates" (1987, 55). These scientists were well aware of the rhetoric that they built into their letter. I also draw your attention to the use of the first person in this important example of scientists' writing.

Brian Martin outlines how scientists selectively present their methods and results to support their conclusions and use (carefully embedded) emotive language to persuade. He thinks that scientific authors "do not make balanced analysis, and do not present results in a neutral manner," and they do push a line (1979, 25). He contrasts two papers (one published in *Science*, the other in *Nature*) on gaseous emissions and the ozone layer to show how the choice of words and turns of phrase indicate the underlying ideology of the authors: one calls for action on environmental degradation, the other defends the economics of current practices. He notes, for example, "ozone shield" compared with "ozone layer"; "permitting the harsh radiation" compared with "increasing the ultraviolet radiation reaching the planetary surface"; "burden of NO" compared with "amounts of NO"; "threats to stratospheric ozone" compared with "interact with, and so attenuate" ozone levels (Martin 1979, 36). This is but one example of how scientists use words to persuade through triggering emotional responses in their readers.

Bruno Latour agrees that the distinction between scientific literature "and the rest, is not a natural boundary" (1987, 62). Latour says it is merely the complexities of the way the social purposes of the author are realized that make the scientific paper appear different. When discussing the writings of "science in the making" Latour says: "the more we get into the niceties of scientific literature, the more extraordinary it becomes. It is now a real opera" (1987, 53). The "cast of characters" includes all the cited authorities, the "crowds of people mobilised by the references," (1987, 53). The props are the experimental tools, the heroes are the authors, the villains are the problems encountered and solved.

More than thirty years since Medawar's exposé, the scientific paper is still taken as the definitive piece of scientific writing: the standard. Yet there are many other styles of writing produced by scientists: letters (to each other and to journals), posters at conferences, jottings and unformed notes on their own ideas and other people's work, email messages, grant applications, and textbooks. Through writing in multiple genres, scientists illustrate that the complexities of so-called scientific writing are great and that any attribution of a unitary set of practices is misleading. Despite such deconstructions of scientific writing, there are linguists (e.g., Michael Halliday and Jim Martin 1993; Charles Bazerman 1988) who still analyze it in such a way that they construct scientific writing—as they perceive it—as worthy of holding a "privileged position" (Locke 1992, viii) for all communication in and about science.

Writing forms Science

Writing *forms* (shapes, defines) science because what is published provides the parameters, the paradigms, and hence the possibilities within which scientists work. Frances Bacon, a nonscientist, wrote a science fiction story, called "The New Atlantis", about an ideal society on a remote island (Brown 1986). In it Bacon claimed superiority for empirical science's knowledge production process based on its supposed method—the use of direct observation and experimentation and the induction of general laws—and its powers of prediction and technological control.[5] Bacon's hypothetical scientific method is still described in many introductory science textbooks today, as if it represents practice. I draw your attention to the underlying metaphor relating to an empire embedded within the concept of "empirical" itself, clearly signaling and claiming power for such an interpretation of science. "The New Atlantis" is an example of a piece of (fictional) writing that still forms the practice of science. It does so because many scientists actually perceive that they do proceed by induction, uncovering the pre-existing laws of nature as described in Bacon's idealistic method. But "paradoxically, the Baconian view of science is not based on empirical observation of what scientists do" (Max Charlesworth 1982, 8).

Particular writings form fields of research, through reinforcing "normal" science or by creating a Kuhnian revolution and paradigm shift. For example, the Watson and Crick (1953) DNA letter shaped a new field, giving it direction and purpose: trying to manipulate DNA's "copying mechanism."

Bacon's idealized model of a scientific method has ultimately led to the construction of positivist Western science as a major regime of truth in our society. The highly valorized position that hegemonic science holds has led to the creation of a mystique surrounding it, parallel to that which previously surrounded the hegemonic truth generators of previous centuries: mythology and theology. Richard Lewantin, esteemed biologist, coined the phrase "physics envy" to explain the condition many biologists suffer from when they try to reduce their work to simple cause and effect relationships governed by mathematical laws, like classical physics has done (David King 1996, 39). This reductionism has also had an impact, up until recently, on the methodologies deemed appropriate in the social sciences and the consequent dominance of positivist research designs in education.

In this sense, research in all fields of Western thought has suffered from physics envy with its controlling "positivist gaze" of scientific methodology. I use this term as analogous to the concept of the "male gaze" in cultural studies, a gaze that has the power to shape women's social behavior, because the positivist gaze has shaped all research in Western society while it has been used as the measuring stick against which all other research methodologies have been judged. Hegemonic science thus shapes our view of what counts as knowledge in any field. Bacon's advocation of particular approaches to inquiry as the purest form of truth generation has cast a long positivist shadow not only across science itself, but also across all of the social sciences as well. Two decades ago Ruth Wallsgrove reminded us that "you can't be rational if you pretend that everything you do is rational; if you don't examine and come to terms with what you feel, your feelings will interfere anyway, but in a hidden and uncontrollable way" (1980, 235).

The cast of characters who have deconstructed examples of scientific writing in ways that align with my ideological positioning include Brian Martin (1979), Bruno Latour (1987), Roger Shepard (1988), Jone Rymer (1988), Alan Gross (1990), and David Locke (1992). Each of these authors has not only shown how scientific writing shares techniques with literature and how scientists do write with passionate voices, but also shown how scientific writing forms the very science that it supposedly reports. In their writing, authors' paradigms and ideologies are exposed, their political intentions of persuasion are clear, and their desire for recognition is put on the table. "Scientists are tellers of tales, creative writers who make meaning and who choose the ways they go about doing so" (Jone Rymer 1988, 244).

Writing informs Science

From Isaac Newton's (1687) *Principia*, through Charles Darwin's (1859) *Origin of the Species,* and Albert Einstein's (1905) paper on the "Special Theory of Relativity", to Watson and Crick's (1953) DNA letter, science has been *informed* by the publication of scientists' writing. Through reading others' work, scientists have been able to "stand on the shoulders" of those who came before them (as Newton allegedly said).

Scientists use writing to inform their work, but little of the informing is uncontested (Hildebrand 1996). For example, the competing discourses of religion and science have resulted in court cases and contestations over the school science curriculum as a result of Darwin's work. History tells us that it took thousands of molecular biologists more than forty years to unravel the "possible copying mechanism" so cleverly alluded to by Watson and Crick (1953), and the current social debates surrounding the ethics and dangers of recombinant DNA technologies are an example of people outside the community of practitioners influencing scientists through public critique and challenge—and, perhaps, through control of practice via legal strategies.

Science itself has also always been informed by imagination within writing through the use of metaphorical devices, allegories, and imagery to push out the boundaries. People such as James Clerk Maxwell, Michael Faraday, Friedrich Kekulé, Albert Einstein, and Nikola Tesla have been well documented as sources of inspiration for others (e.g., Roger Shepard 1988; Bernard Dixon 1989, David Locke 1992; Alan Gross 1990; Clive Sutton 1996). Scientists' writing abounds with creative imagery that has led to new ways of thinking about our world.

Cognitive Effect of Writing Imaginatively to Learn Science

> Imaginative writing
> Involves creative thinking,
> Synthesis and active researching.
> Showing that you can
> Appropriate the language
> Of power.
> Pushing the boundaries of genres,
> Combining, blending, bending.
> Showing the teacher more than they knew
> About you
> And your constructions of science.

* * * * *

All of my collaborating teachers claim that the hybrid genres of imaginative writing can support deep and meaningful learning (in contrast to rote and surface learning) when used as either a learning activity or an assessment tool. These teachers argue that imaginative writing tasks can be used to

- identify alternative conceptions;
- learn science concepts;
- synthesize science concepts;
- transform learning into students' words;
- aid memory work;
- and demonstrate learning—to teachers and to students themselves.

Teachers' Claims on the Cognitive Benefits of Using Imaginative Writing

(Voices in the Left Margin)	**Conventional Science Writing**
	. . .
• Identify alternative conceptions: "It's one of the quickest ways to pick up misconceptions	as the object moves on the frictionless

the kids have about anything. You find out if they haven't got a clue about this, or they haven't made the connection between this and that." (Sandy)

- Learn science concepts:

These teachers argue that the act of writing is a powerful learning process—imaginative writing was not seen merely as an exercise to *reveal* prior learning, but also to *provide for* learning.

"I use it because I think a lot of the learning happens when they're doing their story." (Michael)

- Synthesize science concepts:

"I think it helps them make more connections, a lot more connections, and think differently about the information that they've got. Not just put it in a little box and process it as a little chunk of isolated information." (Rosemary)

- Transform learning into students' words:

"It really gets the kids to internalize [ideas] and understand what it's all about . . . They can't fool you in creative writing tasks . . . They have to re-organize it—they can't just take information and transcribe it . . . They have to put it in their own words and re-work ideas. They have to be able to understand the information in their head to be able to write the story." (Sandy)

"Doing it this way, kids can't just get a book or CD-ROM and copy stuff . . . Because, they can find the information all right, but then putting it in their own words, they find really difficult, and I think the creative writing makes them do it in their own words." (Michael)

- Aid memory work:

"It's difficult to learn 'facts' and if they can do it in a fun way they can associate it with something then it stays . . . a couple of kids said to me the other day 'we still remember those element songs from year 7' and they're in year 9 now." (Katie)

"I think they will remember, three years later, the imaginative writing they did, whereas they won't remember from day to day what notes they took from a board." (Nicole)

- Demonstrate learning—to teachers:

"I think they like having other opportunities to show

surface it remains at a . . .

. . .
photosynthesis involves light acting as a catalyst so that
. . .

. . .
the third sample was rendered unusable and it was discarded so that it . . .

. . .
while the red blood cell carries the xygen around othe hemoglobin acts as . . .

. . .
oxidation occurs at the anode in the cell because
. . .

. . .
the common structural

what they understand. Some students find it an easier way to show what they understand than other methods: often kids who hate tests really do well at a creative writing response." (Sandy)

- Demonstrate learning—to students themselves:
"For a start they have to acknowledge they don't understand and I think that's a big thing. That's where the real learning comes in: when they find out what they don't know and they have to go and find out more." (Rosemary)

"It was a real discriminator for them: of the bits they could understand and the bits they were very confused on. I've found it's one of the most effective things I've ever used for kids—for themselves—to really see if they understand what's going on." (Sandy)

characteristics
mean that . . .

. . .
the aperture
generates diffraction
patterns that can
. . .

Playing with Language

Desire
For pleasure in the pursuit
Of learning.

Fun,
In students' words: "not boring,"
Motivates.

Engagement
Is attention focused on ideas:
Flow.

Learning,
Has pleasure at the heart
Of it all.

* * * * *

Many of those who advocate the use of imaginative writing as a feminist practice promote a sense of playfulness in learning through writing (e.g. Leonie Gianello 1988; Sue Lewis and Anne Davies 1988; McClintock Collective 1989; Liz James 1989; Gaell Hildebrand 1989; Sue Stocklmayer 1989). The concept that it is permissible to have fun in a science lesson is not new; neither is the idea of having fun through playing with writing. "It seems likely that play with written language can help children develop greater awareness and understanding of how it can be manipulated—of what can be done with it and of what they can do with it. . . . Playing with the processes and forms of writing seems likely to give children a sense of ownership" (Joan McLane 1990, 312). McLane describes play as allowing students to stretch their learning and feel as though they "were already competent," or, in Lev Vygotsky's words: "Play creates a zone of proximal development of the child. In play a child always behaves beyond his [sic] age, above his daily behaviour; in play it is as though he were a head taller than himself" (Vygotsky 1978, 102).

Some students would be familiar with playfulness in popular books such as Joanna Cole's (1989) "Magic School Bus" series—and its Microsoft multimedia reincarnations on CD-ROM. A teacher who heard a presentation by Cole, as part of a program supporting the integration of science with language arts, reacted: "Joanna Cole represents all that is good about combining science and literature. She exemplifies how naturally they fit together" (Linda Baker and Wendy Saul 1994, 1029). Cole uses hybrid imaginative writing genres. We ought to remember that not only do young children like to play, but so too do secondary students and their teachers. Play amongst science education researchers can trigger fractures and paradigmatic shifts by slipping into the gaps in the dominant discourse of writing to learn science. As Jay Lemke exhorts, "Make trouble. Play!" (1995, 184).

> Part of the pleasure of it for the kids, is that it allows them to have a bit of fun. So they can be imaginative and creative in what they're doing . . . It's not one of those strategies where you have to drag the kids along, kicking and screaming, to get it done . . . They really engage with the task; they really enjoy that opportunity to look at it in another way. (Sandy)

One teacher explained that those students who expect science to be a dull and/or difficult subject are surprised that they enjoy themselves.

> Some kids enjoy it so much that they actually think they're not doing science. It's so different to what they thought that you did in science that this mustn't be science, because this is actually good. (Pat)

Her comment is quite a revealing one and a sad commentary on hegemonic pedagogy in science. A new both/and relationship emerges: imaginative writing means students can *both* learn in a science class *and* have fun.

> Disturbances, fractures and interruptions to science
> Mean thinking beyond the old pedagogies of compliance,
> Remembering those whose science lessons were for nought

> Creating contestations in what and how
> Science could,
> And should,
> Be taught.

* * * * *

Power and Ownership of Scientific Language

> You can't learn properly unless you take on a new language. But I think the trouble with science definitions is not knowing what the words mean, but it's that language style: it's that third person, all pervasive language style that seems to define it . . . Everyone doesn't want to come out of science with restrictions on the boundaries of how you write. I've been amazed how nearly everyone [who teaches science] feels so constrained by the formal language that we learnt in science and we haven't been able to throw it off. It has really felt like a big clamp that has shut you down in how you express yourself. (Chris)

Along with myself there are now many other science and language educators who agree that to use *only* the supposedly scientific-factual genres in learning science is to re/present hegemonic Western science. For example, Clive Sutton advocates that teachers ought to "present the language of science as a human product . . . using an interpretative voice" themselves, putting "everyday expressions and technical language together," and they should also "encourage the interpretative voice" in their students so that "it is clear that language is a medium for conversation about ideas" (1996, 13). If students are to become active in the discourse of power/knowledge, then there must be a gradual shift from where they are now (users of everyday language) to where they may choose to be (immersed in science practice or critique). This is reinforced by Yvette, who uses imaginative writing as a bridge from everyday language into scientific language with her students: "If they don't have the technical expressions they're powerless really. I mean they sound like bunnies, in their language, and I think that they would go on, and they would feel that it was all right to speak about scientific concepts in slang or everyday language."

Sutton concludes that extensive use of scientific language perpetuates an image of detachment of knowledge from people, and he advocates that "ordinary human language, including narratives," would be a better language for learning and for seeing how scientists work (1996, 17). Jay Lemke's (1987) assertion that the use of scientific terminology is interpreted as a "claim to power" and that people come to accept that "superior" people master it is related to the concern of many feminist science educators, a concern that science is directly linked to hegemonic forms of masculinity through, among other means, the discursive practices associated with its learning and teaching (e.g., Nancy Tuana 1989; Sue Rosser 1990; Sandra Harding 1991). The close link between language and learning was clear in all of these teachers' thinking, and some teachers recognized the power that comes with ownership of, and control over, scientific language.

It gives them a sense of power, in a way, that they can actually use that language. (Sandy) Other teachers went further, recognizing that science is directly linked with power through its language and that skill in controlling scientific language can allow some people to manipulate others. As Katie explains, "I think there is a bit of academic snobbery too, in a lot of the so-called scientific language. There's more than just using the scientific language: there's politics and power at play there . . . it's the politics of language . . . You read scientific papers and you sometimes wonder whether or not they're being deliberately obscure. I sometimes think scientists are showing off to each other . . . You can manipulate people and you've got the upper hand, in a sense, if you can manipulate the language of science."

Chris, in recognizing that science is linked to power through its language, warned that those people who allow girls to use simple language, while skilling up boys, are perpetrating a severe injustice: "To use the language of science is to call in power. And we're doing girls a disservice if we don't actually teach the "truth" of how to survive in the science world, and the boys would pick up that jargon and use it, and be powerful in the end, and girls wouldn't. So I think it's important to use both everyday and scientific language. One is a stepping stone to the other."

Directly linked to the power of scientific language is the issue of ownership of it: students who feel they own the language are not intimidated by its power. Providing opportunities for students to feel that scientific language belongs to them is seen as a strength of imaginative writing. As Sam explains, "It's very effective for concepts and processes [as] students are involved in it themselves, it gives them ownership."

Control of the language of science is encouraged by a sense of playfulness: "For the mainstream student, it breaks down the fairly severe image they have of science, the remote image that they have . . . It's the knowledge that it can belong to them, it doesn't have to belong to the people in the laboratories . . . They then feel they can cope with it, and understand it. It's not precious—they can use it. They can feel in control of it. Whereas before, if you heaped this foreign information on them, they didn't want it, they don't want to know about it: it's just too hard. They shut off . . . From what I can understand most kids are not comfortable with science information. They think it is boring and not of their world. This makes it personal and says "it's OK to jiggle this around." They feel that it is so structured that they're not supposed to jiggle it." (Rosemary)

The notion that students can "jiggle" the language of science is not usually found in hegemonic pedagogy. Another way to show students the ways power and language interact is to use Katie's deconstruction model. This highlights the ways science is written about in the popular media and illustrates techniques for students to use: "Those who can manipulate the language of science are in a very powerful position because you can exclude people, and bamboozle them with facts, and whatever. But if people can become more critical of what's being said and how it's being said, perhaps they can then look beyond the obvious" (Katie). Explicitly showing students how others can "bamboozle" them with science is foregrounding to students the way texts work in practice. It is teaching them to "read the world" and not just the word (Joan Wink 1997) from a critical pedagogy position. These teachers are trying to enable their students to critique and challenge the science in their lives so that they will not be intimidated by people like journalists, doctors, and politicians.

When asked if students should write only in the ways that scientists do, Elaine's response was unequivocal: "Clearly not. We're talking about changing the whole world, changing the way science is conceptualized, taught and popularized. I think part of that is breaking into the notions of what writing about science is. I think that some of the writers out of popular literature have taken the creative initiative and are doing all those sorts of things. There are lots of places where it's happening already and we were behind, putting it into the classroom."

The writing genres used for learning science do not need to be congruent with those that scientists use because of the fundamentally different social purposes of the two projects: learning science and reporting science. These teachers worked on the premise that science teaching and learning should *deliberately* utilize different language practices, including a wide variety of writing forms, because of the accessibility of the conceptual development that they provide. As Katie describes, "There's a vast difference between what you're expecting people to produce [in science] later on, and what kids are interested in, in years 7 and 8. I think you would be doing the kids a disservice if you had a particular thing in mind—this is the scientific way of writing, "this is the best way" . . . If we try and sausage kids into this idea that you *only* write reports for science, and that's the way that you learn science, then it's really narrowing and I think that if you decide that's the way to go, you'd be turning girls off."

As Elaine said, teachers "transmit the culture of science" through their classroom practices. Richard Coe, a linguist, argues that "genres are . . . important factors in the social construction of orientations, paradigms, ideologies, world views, and cultural perspectives" (1994, 184). He says that genres can be seen as "fossilized rhetorical processes" and for them to work they must "achieve desired effects" (184). I agree with Coe; the "tyranny of genre" exists when there is an unreflective and formulaic application of narrow genres. Coe sees a danger here because to "help someone learn a generic form," such as a laboratory report, without also helping them become critically aware of its effect, "may subject them to the genre rather than empower them" (1994, 188).

For me this means that in school science we need to be reflective on what the effects of particular genres might be. We need to ask questions of ourselves: What are the social purposes of writing in science classes? Why do we insist on laboratory reports being written in this particular style? What other writing forms might better achieve the effect we desire? Aviva Freedman and Peter Medway argue that there are two grounds for reappraising the ways genres have been advocated in schooling. First, genres do not simply "display the disciplinary knowledge" of a subject like science, but also "enact the social relations of science" in classrooms—relations that have served a "filtering function that systematically denied legitimacy" for "females or working class students" in the past (Freedman and Medway 1994, 14). "The perpetuation of the genre contributes to the perpetuation of those forms of power and exclusion" (14).

Patricia Rowell concludes that "writing in school science could be used to move in multiple directions among discourses" rather than constraining students to write within set genres all the time (1997, 47). She argues that "if multiple roles for writing in science are recognized, the taken-for-granted communication function, tied to transmission of information . . . could be dislodged from its predominant position" (1997, 48).

By explicating the purposes of conventional writing styles and genres, and providing models beyond these, the teachers reported here are challenging the hegemony of conventional writing practices. They recognize that boundary policing of the right way to write science serves to marginalize many students beyond the borders of science.

> In theorizing theory,
> I challenge its duplicity

In maintaining the sanctity,
And hegemony,
Of science and a didactic pedagogy.

In promoting critical activity,
Utilizing the affective and creativity,
A new "enabling" pedagogy
Challenges the authority
Of discourses that assume priority.

Imaginative writing re/presents science and pedagogy
As contestations to hegemony.

* * * * *

Creativity in Science

These teachers hold a belief that science is a creative process and that by encouraging creativity they are making an important contribution to developing problem-solving skills.

> Admittedly the practice of science sometimes can be tedious, but you're looking at people who are ultimately trying to solve problems. And you're not going to solve a problem unless you're going to be creative and innovative about it, in some way. (Sandy)

Laura argues strongly that science should not be portrayed as a catalogue of facts but as a dynamic field changed by creativity: "It gets them thinking more creatively, rather than looking at science as a fact-finding type of subject or discipline. They come to realise that a lot of hypotheses are not necessarily "fact" for long, they're always being revamped. To believe that science is just so wholesome and truthful and without any creativity is, I think [pause] that's a *sin!*" (Laura). Laura's strength of view is evident in her language. Alex also links training in creativity to the dynamic nature of science, expressing a desire that one day creative students, like hers, will be able to change science: "The way we think about things now may not be the way we think about them in the future: and possibly that might happen because people are encouraged to think about them differently, in school."

In trying to live their philosophy of science teaching and learning, these teachers attempt to innovate and vary their teaching strategies: "We try and teach science creatively, as far as possible" (Katie).

Overcoming the noncreative stereotype of science and its teachers is a concern of some of these teachers—not only with their students but also with their colleagues. Angela's comment highlights the notion that teachers who use imaginative writing are creative and resourceful themselves: "This is my whinge, OK. We still have English Literature people here at school, who think that we all belong in a box over on this side of the school, who only think of E=mc². And that really is upsetting because most of us are pretty well read in many things . . . We all bounce pretty good ideas off each other when we have a meeting and we all have ideas to make it creative. Because we're NOT the boring types in the labs any more." One downside that Katie mentioned was this: "It gets to the stage where they just don't want to listen to "chalk and talk" any more." As "chalk and talk" has long been associated with hegemonic pedagogy, it is not surprising to me that these students reject it once they've experienced creative aspects of a more enabling pedagogy (Hildebrand 1998).

Changing Pedagogy

> Science teachers
> Thinking ideologically
> While acting practically
> Re/constructing a pedagogy that valorizes,
> Moving towards one that problematizes
> Science.

* * * * *

As Walter Parker and Janet McDaniel (1992) argue, teachers are "bricoleurs" who are "masters of improvisation" transforming others' strategies in order to make them work within their own context. When Michael reflected on his initial resistance to the idea of using imaginative writing, he advised teachers to try it before being too judgmental:

> My first reaction to creative writing was probably the same as a lot of other people, you say, 'oh, what a load of rubbish' . . . [But] I'm pretty sure that every science teacher should at least tackle it a couple of times and just see how it works . . . it's endless, the variety of things and the techniques you can use . . . It's really changed my thinking—from a traditional science teacher—to realise that there are other ways that you can achieve results and maybe in a more effective way. (Michael)

Katie, a true bricoleur, recalls that once she tried imaginative writing she permanently built it into her repertoire.

> I've always been interested in how kids learn. If you really want to do your job well you're always on the lookout for new things and how you can be most effective as an educator. And I've always been pretty well committed to what works well in education—and this does. (Katie)

I conclude that writing *informs* science and science learning: that is, it illuminates or provides new insights for scientists, teachers, and learners. In science learning, imaginative writing could generate new ways of thinking in and about science, and new ways of learning for marginalized students. "Forms of classroom writing may be sanctioned that reflect the language, value-orientations, and relevances that students bring into the classroom with them, instead of [only] ones modeled on discourses of the discipline community" (Freedman and Medway 1994, 15).

In this chapter I have opened up clear spaces for dialogue and provided possibilities for disruption of the discourse on the right way to write science, especially in schools. I think that to uncritically perpetuate writing practices that are implicitly underpinned by an ideology that links science with power (and masculinity) is to choose to teach in ways that generate privilege for *some* students.

Those of us in the margins can re/write school science and, perhaps, science.

* * * * *

The coalesced idea explored
And presented in this collation
Is
Writing to learn science
Can fruitfully engage the imagination.

References

Atwater, M. M. 1994. Research on cultural diversity in the classroom. In *Handbook of Research on Science Teaching and Learning*, ed. D. L. Gabel. New York: Macmillan.

———. 1996. Social constructivism: Infusion into the multicultural science education research agenda. *Journal of Research in Science Teaching* 33: 821–37.

Avery, S., M. Kennard, S. Turnball, S. Denver, M. Crook, J. Devlin, S. Guthry and J. Munro. 1996. Creating an elemental experience. *Lab Talk* 40: 24–25.

Baker, L., and W. Saul. 1994. Considering science and language arts connections: a study of teacher cognition. *Journal of Research in Science Teaching* 31: 1023–37.

Ball, S. J. 1993. What is policy? Texts, trajectories, and toolboxes. *Discourse: The Australian Journal of Educational Studies* 13: 10–17.

Bazerman, C. 1988. *Shaping written knowledge: The genre and activity of the experimental article in science*. Madison: University of Wisconsin Press.

Bazler, J. and D. Simonis. 1989. Gender analysis of chemistry textbooks. In *Gender and Science and Technology (GASAT 5) Proceedings*, eds. I. Ravina and Y. Rom. 125–33. Haifa, Israel: Technion – Israel Institute of Technology.

Brown, H. 1986. *The wisdom of science: Its relevance to culture and religion*. Cambridge, UK: Cambridge University Press.

Charlesworth, M. 1982. *Science, non-science and pseudo-science*. Geelong, Australia: Deakin University Press.

Coe, R. M. 1994. An arousing and fulfilment of desires: The rhetoric of genre in the process era—and beyond. In *Genre and the new rhetoric*, eds. A. Freedman and P. Medway. London: Taylor and Francis.

Cole, J. 1989. *The magic school bus inside the human body*. New York: Scholastic.

Crick, F. 1988. *What mad pursuit: A personal view of scientific discovery.* London, UK: Penguin.

Dixon, B., ed. 1989. *From creation to chaos: Classic writings in science.* London, UK: Basil Blackwell.

Esterhuysen, P. 1996. 'Focusing on the frames': Using comic books to challenge dominant literacies in South Africa. In *Challenging ways of knowing in English, maths, and science*, ed. D. Baker, J. Clay, and C. Fox. London: Falmer Press.

Freedman, A., and P. Medway, eds. 1994. *Learning and teaching genre.* Portsmouth, NH: Boynton/Cook.

Freire, P. 1968. *Pedagogy of the oppressed.* New York: Seabury Press.

———. 1985. *The politics of education: Culture, power, and liberation.* South Hadley, MA: Bergin and Garvey.

Gianello, L., ed. 1988. *Getting into gear: Gender inclusive teaching strategies in science developed by the McClintock Collective.* Canberra, Australia: Curriculum Development Centre.

Gross, A. G. 1990. *The rhetoric of science.* Cambridge, MA: Harvard University Press.

Haggerty, S. M. 1995. Gender and teacher development. *International Journal of Science Education* 17: 1–15.

Halliday, M. A. K. 1993. Some grammatical problems in scientific English. In *Writing science: Literacy and discursive power*, ed. M. A. K. Halliday and J. R. Martin. London: Falmer Press.

Halliday, M. A. K., and Martin, J. R. ed. 1993. *Writing science: Literacy and discursive power.* London: Falmer Press.

Hand, B., and V. Prain, 1996. Writing for learning in science: A model for using in classrooms. *Australian Science Teachers' Journal* 42: 23–27.

Harding, S. 1991. *Whose science? Whose knowledge?* London, UK: Open University Press.

Hart, C. 1995. *Access and the quality of learning: The story of a curriculum document for school physics.* Ph.D. Diss., Monash University, Australia.

Hernández, A. 1997. *Pedagogy, democracy, and feminism: Rethinking the public sphere.* Albany: State University of New York Press.

Hewlett, L. 1996. "How can you 'discuss' alone?": Academic literacy in a South African context. In *Challenging ways of knowing in English, maths, and science,* ed. D. Baker, J. Clay, and C. Fox. London, UK: Falmer Press.

Hewson, P. W. 1981. A conceptual change approach to learning science. *European Journal of Science Education* 3: 383–96.

———. 1996. Teaching for conceptual change. In *Improving teaching and learning in science and mathematics,* ed. D. F. Treagust, R. Duit, and B. J. Fraser. New York: Teachers College Press.

Hildebrand, G. M. 1989. Creating a gender inclusive science education. *Australian Science Teachers Journal* 35: 7–16.

———. 1991. Using non-conventional writing genres in science. In *Proceedings of the WA Science Education Research Association's annual conference,* ed. M. Hackling, 61–73. Perth, Australia: Edith Cowan University.

———. 1996. Writing in/forms science learning. Paper presented at the Annual Meeting of the National Association for Research in Science Teaching, April, St. Louis.

———. 1998. Disrupting hegemonic writing practices in school science: Contesting the right way to write. *Journal of Research in Science Teaching* 35 (4): 345–62.

James, L. 1989. The atoms' party. *Australian Science Teachers' Journal* 35: 72.

King, D. 1996. Biology's best known rebel won't wear genes. *The Australian: Higher Education Supplement* (June 26): 39.

Kuhn, T. S. 1970. *The structure of scientific revolutions* 2nd ed. Chicago: Chicago University Press.

Latour, B. 1987. *Science in action.* Cambridge, MA: Harvard University Press.

Layton, D. 1994. STS in the school curriculum: A movement overtaken by history? In *STS education: International perspectives on reform*, ed. J. Solomon and G. Aikenhead. 32–44. New York: Teachers College Press.

Lemke, J. 1987. Talking science: Content, conflict, and semantics. Paper presented at the Annual Meeting of the American Educational Research Association, Washington, DC. ERIC Document No. ED 282402.

———. 1995. *Textual politics: Discourse and social dynamics.* Chicago: Taylor and Francis.

Lewis, S., and A. Davies. 1988. *GAMAST professional development manual: Gender equity in mathematics and science.* Canberra, Australia: Curriculum Development Centre.

Locke, D. 1992. *Science as writing.* New Haven: Yale University Press.

Martin, B. 1979. *The bias of science.* Canberra, Australia: Southwood Press.

Martin, J. R. 1996. A girls' pedagogy in relationship. In *Equity in the classroom: Towards effective pedagogy for girls and boys,* In eds. P. M. Murphy and C. V. Gipps. London, UK: Falmer Press.

McClintock Collective. 1989. Creative writing ideas. *Australian Science Teachers' Journal* 35: 72–73.

McLane, J. B. 1990. Writing as a social process. In *Vygotsky and education: Instructional implications and applications of sociohistorical psychology,* ed. L. C. Moll. Cambridge, UK: Cambridge University Press.

Medawar, P. 1963. Is the scientific paper a fraud? *The Listener, BBC Publications,* (Sept. 12): 377–78. Reprinted in *From creation to chaos,* ed. B. Dixon, 170–75. London, UK: Basil Blackwell, 1989.

Parker, W. C., and J. E. McDaniel. 1992. Bricolage: Teachers do it daily. In *Teacher personal theorizing: Connecting curriculum practice, theory, and research,* ed. E.W. Ross, J. W. Cornett, and G. McCutcheon. Albany: State University of New York.

Prain, V. 1995. Writing for learning in science. In *Teaching and learning in science: The constructivist classroom,* ed. B. Hand and V. Prain. Marrickville, NSW: Harcourt Brace.

Rosser, S. V. 1990. *Female friendly science: Applying women's studies theories and methods to attract students.* New York: Pergamon.

Rowell, P. M. 1996. The images of science in the genre debate. *Australian Science Teachers Journal* 42: 29–33.

———. 1997. Learning in school science: The promises and practices of writing. *Studies in Science Education* 30: 19–56.

Rymer, J. 1988. Scientific composing processes: How eminent scientists write journal articles. In *Advances in writing research, Vol. 2,* ed. D. A. Jolliffe. Norwood, NJ: Ablex.

Shepard, R. 1988. The imagination of the scientist. In *Imagination and Education,* ed. K. Egan and D. Nadaner. Milton Keynes, UK: Open University Press.

Snow, C. P. 1963. *The two cultures and a second look.* Cambridge, UK: Cambridge University Press.

Stanley, W. B., and N. W. Brickhouse. 1994. Multiculturalism, universalism, and science education. *Science Education* 78: 387–98.

Stocklmayer, S. 1989. Creativity and girls in science. *Australian Science Teachers' Journal* 35: 77.

Stronach, I., and M. MacLure. 1997. *Educational research undone: The postmodern embrace.* Buckingham, UK: Open University Press.

Sutton, C. 1992. *Words, science, and learning.* Buckingham, UK: Open University Press.

———. 1996. Beliefs about science and beliefs about language. *International Journal of Science Education* 18: 1–18.

Tuana, N., ed. 1989. *Feminism and science.* Bloomington: Indiana University Press.

Vygotsky, L. S. 1978. *Mind in society: The development of higher psychological processes.* Cambridge, MA: Harvard University Press.

Wallsgrove, R. 1980. The masculine face of science. In *Alice through the microscope,* ed. Brighton Women and Science Group. London, UK: Virago.

Watson, J. D., and F. H. C. Crick. 1953. Molecular structure of nucleic acids: A structure for deoxyribose nucleic acid. *Nature* 171: 737–38.

Watts, M., and D. Bentley. 1994. Humanising and feminising school science: Reviving anthropomorphic and animistic thinking in constructivist science education. *International Journal of Science Education* 16: 83–97.

Wink, J. 1997. *Critical pedagogy: Notes from the real world.* White Plains, NY: Longman.

Notes

1. Like Christina Hart (1995) I have reported the first names of authors to challenge the conventional academic referencing style that depersonalizes and disembodies in order to abstract ideas from the murky pool of subjectivity and to ascribe a higher level of objectivity and authority to published work. First names signify that all arguments and theories are constructed by real people and remind us that we all perceive our world through individual frames of reference that represent our multiple subjectivities.

2. Re/present. I have used the "mysterious slash", a tool from poststructuralism, that "both joins and separates like a scar" (Ian Stronach and Maggie MacLure, 1997) to draw attention to boundaries, spaces, connections, contradictions, and multiple meanings. In this case re/present can be read as represent, present again, present in an alternative way, and present from another frame of reference.

3. The collaborating teachers, whose voices are indicated in this chapter by a pseudonym, were part of a study that I undertook with twenty exemplar teachers. They had each used imaginative writing as a learning and/or assessment tool for at least three years prior to the teacher and student interviews, classroom observations, and student questionnaires of the study that informs this chapter.

4. Gaell is a member of the McClintock Collective, a gender-science-education network that has run many professional development activities on inclusive teaching. She was a member of the team who ran the workshops that Michael attended.

5. The metaphors used by Bacon laid down the foundations of the discourse patterns that still construct science as a gendered endeavor.

Chapter 8

Rethinking Science and Assessment

Dana Fusco

> These people came out here today because they love the people in the community and themselves. They want to make this community a better place, even though they won't live here for a long time.
>
> —Ty, age thirteen

> I used to live in the shelter. I wanted to come and help the community to make it a better place for kids to come and play. Instead of being out in the street and doing things they shouldn't be doing, they just come in here and relax and enjoy themselves.
>
> —Mia, age fifteen

These voices were expressed on Community Day, the day that Ty, Mia, and other members of REAL put their environmental design plan into action. REAL, or Restoring Environments and Landscapes, was the teenagers' self-defined name for the group and reflects a vision for science education that is inclusive, meaningful, and relevant to the lives of young people. Over the course of nine months, I worked with homeless youth designing and implementing a community-based science project and assessment. Our goal, which emerged gradually, was to transform an empty lot across the street from the shelter, where they temporarily resided, into a usable public space for the community. To accomplish this goal, a practice was created that included science, mathematics, technology, photography, journalism, art, research, environmental design, and more. The activities emerged from what was interesting and necessary to pursue towards the realization of our goal(s); the assessment emerged alongside practice. Was REAL "real" science? What is transformed when science is created with what young people bring to their learning environment? How do such practices inform how we think about knowing in science, learning, and assessment?

Through a historical narrative of my work with urban teenagers, I will describe the participatory planning and design process that allowed the young people to produce science in the service of community and present how an assessment tool emerged alongside this practice.[1] I begin this chapter by bridging reform efforts in science education with assessment reform. Knowing and learning in science, as well as assessment in science, are no longer understood as methods for discovering neutral and singular truths but are situated in social, cultural, historical, and political contexts and in the perspectives of learners and teachers as well as in the tools available within the learning environment, including the assessment itself. In describing my practice with urban youth, I highlight three features of knowing and learning in science, features that were made visible through the content and form of the assessment: knowledge as borrowed and reinvested, scientific plausibility in the context of local action, and knowledge as public and publicly enacted. In the concluding section, I will revisit reform objectives in science and assessment, allowing this practice to bump up against theory and research. The results reflect how science knowledge was emergent, recursive, and situated in cultural and historical practice and suggest the need for assessments to be emergent also. In short, I hope to demonstrate the process of creating an inclusive science practice, the conditions necessary for its existence, and what that implies about science knowledge, the purpose of engaging in science-related activities, and science assessment.

Bridging Reforms in Assessment and Science Education

In 1990, the American Association for the Advancement of Science proposed the creation of a new and inclusive model of science education. Scientific literacy under this new paradigm includes "knowing that science, mathematics, and technology are human enterprises, knowing what that implies about their strengths and limitations; and being able to use scientific knowledge and ways of thinking for personal and social purposes" (AAAS, 1990, xvii–xviii). These goals coincide with multicultural, critical, and feminist traditions that view science knowledge as developed for particular social and cultural purposes. However, they are also antithetical to multicultural, critical, and feminist visions for science education because they do not situate knowing in science within political and historical contexts and therefore do not address the barriers that have served as obstacles to equity and excellence in the past. From a critical science perspective, inclusive and equitable practices in science begin by questioning the nature of science, who one must be to participate in it, and the methods by which scientific knowledge claims are made (Barton 1998a; Eisenhart, Finkel, and Marion 1996; McShane and Yager 1996; Rodriguez 1997, 1998). *Critical science* informs the dialogue on the limitations of science and scientific methodologies, i.e., that the knowledge derived through scientific means is not absolute, offers one perspective, and is situated in sociocultural and political contexts. It critically examines what constitutes legitimate science knowledge and suggests the need in education to create spaces where different voices not only can be "heard" but can participate in creating an inclusive science. As Barton argues in her book *Feminist Science Education*, "the role of science class was not simply to help the students 'do science' but rather to do that which grows out of their questions and experiences" (1998, 112). Efforts to expand the boundaries of science and pedagogy in science come on the heels of critiques of school science as being narrow in scope and excluding the lived experiences of many young people (McShane and Yager 1996). Critical science seeks to connect science to individuals and society in meaningful and relevant ways by engaging participants in the production of science towards local action (Ahlgren and Rutherford 1993; Barton 1998a, 1998b; Eisenhart, Finkel, and Marion 1996).

In this chapter, I hope to add to the dialogue a vision for science assessment that is also inclusive and emergent from the lived experiences, interests, and concerns of young people. This vision is in part grounded in critiques of the objective and often alienating nature of educational testing as well as responses to such critiques. Namely, standardized tests have been criticized as reflecting a particular way of knowing and not accurately assessing what *all* people know in science (Kulm and Stuessy 1991). Many tests assume sequential mastery of discrete facts and skills (Shepard 1991) and omit the assessment of critical thinking and reasoning (Nickerson 1989). By informing students of their abilities, such assessment procedures have often alienated girls and minorities, leaving them ill-equipped to handle the work and less interested and motivated to pursue scientific endeavors (Roth and McGinn 1998).

Over the last decade, alternative methods of assessment have been constructed to provide more equitable measures of learning as well as additional opportunities for "using a mind well." The use of alternative assessments was sparked by the 1988 educational reform in Great Britain, "touted as model educational policy" (Baker 1994, 453). Educational reform in the United States has come to reflect the assumption that high national standards and expectations of excellence for all Americans, as well as fairer assessments, would level the playing fields in mathematics, science, and technology (Malcom 1991). In 1992, a congressional committee, the National Council on Education Standards and Testing (NCEST), recommended the increased, but cautious, use of performance-based assessments linked to high national standards. Performance-based assessments are in vivo assessments of actual performance and show what a student can do rather than what a student can skillfully recover from memory. They are embedded in complex, authentic situations both assessing situated reasoning and creating further opportunities for problem solving (Herman, Aschbacher, and Winters 1992; Wolf et al 1991). For example, students might design an experiment testing the effects of various factors on plant growth and communicate their results to an audience. The assessment creates further opportunities to do and talk science as students demonstrate their knowledge of life systems. Learning is no longer viewed as the acquisition of discrete information but seen as a complex, socially distributed process of constructing meaning. Because performance assessments do not require the recall of discrete bits of knowledge, disadvantages for females and language minority students may disappear (Ruiz-Primo and Shavelson 1996), particularly when students have a choice over the direction of the task and the response format (Baker, O'Neil, and Linn 1993).

The very nature of performance-based assessment challenges our understanding of science and of assessment in science by situating scientific inquiry in a socially constructed context. Performance assessments hold the potential to transform the nature of science and pedagogy in science from one that is fact focused and contextually abstract to one that is emergent from the life experiences of young people. Science for "personal and social purposes" then might be inclusive of local dialogues and experiences. The assessment, rather than deleting the achievements of marginalized groups in science, would reflect and help create equitable practices in science education by virtue of its embeddedness in the co-production of local science. That is, the assessment might also grow organically from the inquiries and actions in which teachers and students are engaged. As I will show below, the questions and actions that arise in response to doing science in the community would be at once science and assessment in science.

All of Science

Creating a "science for all," from my perspective, means including my social scientific methods within the discourse of this objective. My approach with the young people can be described as participatory action research (Noffke 1997). Action research challenges the dualistic separations inherent in Western science, separations such as objective and subjective, researcher and researched, and identifies the hierarchical and elitist power structures that such dualisms have created. For instance, through its objective lens, science has created a way of seeing African American children and families as disadvantaged and at risk (Ladson-Billings 1994). While seemingly harmless to the relationship between astronomers and the stars, the dualistic separation between object and subject has created oppressive situations when used to study human beings. It is from this vantage point that I entered my relationships with the young people not to learn something *about* them (and how they understand/use science) but in order to create something *with* them. In fact, I could not learn anything about them without creating something with them. I believe that creating *with* young people requires flexibility in one's approach. Therefore, rather than fixing a method in place, the methodology allowed the researcher and young people to continue to grow simultaneously with their practice (by method, I mean all science activities-educational, evaluative, and empirical). That is, method was not created beforehand and then put into practice; method was created in practice. It was developed in the context of particular relationships, situations, and interests. As Holzman describes it, "practicing method is an explicitly participatory activity that entails the continuous, self-conscious deconstruction of the hierarchical arrangements of learning, teaching, and knowing" (1997, 11). My role was not to place limitations on the particular content area to be taught, learned, and assessed but to support the opportunity for young people to step into new roles as urban planners, designers, researchers, activists, scientists, mathematicians, photographers, and journalists-roles that continuously emerged in the process of developing an inclusive practice-and to direct the group in *how* we might learn, work as a team, and keep track of what we were doing together. It was through performing these roles that young people not only learned *about* design, but were designers. It was in their performances that the assessment tool emerged as a method for documenting this collective practice.

The data includes this documentation as well as my field notes. The field notes were written immediately after each meeting or within twenty-four hours and took the form of a personalized and detailed account of each session. Statements made by the young people or staff were included in the field notes; however, since tape recorders were seldom used, quotes do not necessarily represent verbatim statements but reconstructed ones. My field notes were a valuable source of information, filling in the voids inherent in viewing only end products, helping to interpret the data historically, and highlighting the decision- making processes involved in the implementation of the program. The data was interpreted through a technique of identifying themes and generating ideas (Bogdan and Biklin 1992). This method of qualitative data analysis is an up-from-the-bottom process of examining the data through and through until "it speaks to you." However, the results are not presented and interpreted from the perspective of having discovered some truths about urban homeless children or from the perspective of having found the right pedagogical position from which to teach and interact with homeless children. Rather, the research is presented from the perspective of my ongoing reflections in the practice of simultaneously developing/rethinking science and assessment. Below I will interweave science and assessment as they co-emerged in history, bringing into focus the particular activities of Mia and Ty, two members of REAL.

Organizational Context

Beginning in the fall of 1998, I began working with a group of teenagers with the broad goal of creating a community-based science project. The project occurred twice a week and in collaboration with an after-school program operating out of a temporary housing shelter for homeless families. The shelter, located in an inner-city neighborhood in the Northeast, served over two hundred families. After-school and Saturday services were provided for children and youth in these areas: educational enhancement, social recreation, sports and fitness, cultural enrichment, and teen services. The project worked in collaboration with teen services, geared for youth ages twelve and up. However, because of the nature of the project, younger children often could and did participate. In fact, I did not discourage younger children from participating but supported creating multiple zones of development by allowing heterogeneous learning environments to form (Vygotsky 1978). Some younger children sporadically participated; others became core members of the project.

Over a nine-month period, approximately forty children and teenagers were involved in the project, about one-third of whom were core participants, or attended the sessions regularly. Many of the participants were boys (approximately 68 percent) since fewer teenage girls lived in the shelter at the time and several had other responsibilities, such as attending church or babysitting. Weekend social events, such as Community Day, showed the largest attendance of teenage girls, who often came with their families. The participants were predominantly African American, Caribbean American, Latino, and biracial.

Mia and Ty

Mia, age fifteen, is of West African descent and lived in the shelter for almost a year with her twin brother (also a member of REAL) and both of her parents. Her parents were very interested in her educational achievement, checking that homework was completed each day. Mia was a quiet girl and keenly observant. She excelled in school and demonstrated a responsible and caring work ethic. In fact, during the time that she participated in REAL, she was trained for a junior staff position within the after-school program. Mia described her participation in REAL as motivated by an interest to "help the community to make it a better place for kids to come and play." Mia moved from the shelter during the time of the project; however, she continued to participate in the program.

Ty, age thirteen, was also a core member of REAL. He had been living in the shelter for close to two years with his mother and three-year-old brother. Ty never spoke of his father but often made reference to an uncle, a carpenter that he thought could assist us in building a new fence for the garden. Ty did not do well in school and received mostly D's and F's in his classes. He had a reputation among the after-school and security staff as an "instigator" and "trouble maker." Within REAL, however, he was usually on-task and often provided leadership to the group. In fact, his mother was thrilled and somewhat surprised when she was able to view her son's accomplishments within the project, and she became involved during Community Day events.

Creating the Environment

Action research projects with young people have various agendas; however, most agree with the basic tenet that scientific research should be conducted for the purpose of initiating social change and action. During my first week with the teenagers, I began by asking them, What are the concerns of young people today? They spoke of teen pregnancy, being shot or "making it to the next day," AIDS, unprotected sex, gangs, alcohol use, and adult's perceptions of youth, described as negative "especially if you're black." How could we help to transform the conditions that lead to adolescents' involvement in dangerous activities? I shared with the teens various models of youth-led action research projects. The idea of doing something for/within the community where they lived emerged from the following exchange.

> "Let's have a bake sale and raise some money."
> "Yeah. We could give it to charity."
> "We are charity."
> Silence.

With further discussions about their concerns, we formed our mission-to transform the empty lot across the street from the shelter into a usable community space. The lot sat on a corner surrounded by a chain-link fence that had been knocked down in a "high-speed cop chase," leaving the fence dangling, torn, and unsafe. The lot was also filled with garbage, crack vials, and was both an eyesore in the community and an unhealthy, "smelly" environment. In our mission to restore this urban environment, the teens began brainstorming possibilities for the empty lot: basketball court, swimming pool, arcade, playground, sandbox, garden, stage, cyber games, laser challenge, and a penny store. They conducted a site assessment to determine the feasibility of their ideas. Was the space large enough to house all of these suggestions? What existed in the lot currently and what was its history? Was the soil viable for planting? Four teams were developed in order to measure the space, record its contents (living and nonliving things), take photographs, and draw the present condition of the space. With much excitement, several young people, Mia and Ty included, finished their reports and drawings at home and reported their findings to the group the following week.

Mia was a member of the measurement team and spent hours writing an elaborate report explaining how her team arrived at the perimeter and area of the lot. She explained how there were two methods for obtaining the perimeter: adding the length twice with the width twice or multiplying the length by two and the width by two and adding their sums. Ty was a member of the drawing team. He drew before and after sketches of the lot. His slogan was "We need to live better than this. Please help our environment." In the post-sketch he wrote, "We live better then before." Socially responsible science is defined as "a sense of science as something that is important in their lives and their community outside of school" (Eisenhart, Finkel, and Marion 1996). Here, the study of an urban environment included mathematics as well as conceptual drawings; individuals were not systematically involved in the same process but engaged in different activities that contributed to the whole. How could an assessment capture the complexity of these real-world activities, and what would it say about science and science understanding?

The Emergence of "the Book"

In constructing an assessment tool, I was interested in documenting the process by which our collective activities emerged, as well as the products that resulted. I wanted to create a methodology that captured as well as facilitated our ongoing relational activity. While I had been keeping a collection of our work (photos, drawings, etc.), I formally introduced the idea of documentation to the group as a way to "teach other people about what it means to take an empty lot and turn it into something that's useful for the community." The young people suggested keeping a photo album, "jotting" things down, making a timeline, and creating "a lesson plan . . . like school." The assessment tool emerged from these discussions. Each week new artifacts were added to a three-ring binder. "The book," as the young people referred to it, included actual products (letters, notes, flyers, drawings, etc.), visual representations (photographs), direct inquiries (obtained through surveys, written evaluations/reflections), an attendance log, and a weekly summary of activities (or lesson plans). The book became a documentation of our collective history, and the members of REAL took responsibility for "writing" this history. Some took attendance; some took notes during presentations; some put photographs and other artifacts in the book. At times entries in the book emerged for the explicit purpose of keeping track and "helping us to remember" things we were learning and doing together. At other times entries were actual products, such as Ty's conceptual drawings or Mia's measurement report.

The documentation of REAL's history raises questions about whose knowledge of science is assessed and when. In scientific communities problems are ill defined and require collective efforts. Individual contributions towards a collective product are difficult to isolate. The implication for assessment is that work collectively produced should be understood in relation to what and how the group produced it. Ty's post-sketch of a laser park and an arcade reflects not merely his individual creativity but the ideas for the lot that emerged within the group. As pages in the course of history, these artifacts represent individual talents/strengths and the situatedness of achievements. In a *practice of collaboration* individual contributions are interwoven in the ongoing process and product. Learning is often recursive, continuously getting reinvested into the group, and the endpoints of learning are not so clearly discernible. Ty's drawings and Mia's measurement report were not ends but an ongoing part of REAL's history. Based on the findings of the site measurement, for instance, the size of the lot was judged too small for a basketball court. The teens' initial list was reduced to seven possibilities (playground, garden, clubhouse, penny store, jungle gym, sandbox, and stage). Mia's contribution to this process is seen in history where the "book" allows for these gradual changes over time to be represented. That is, the original list, the teens' reports and criteria for exclusion, and the revised list of possibilities are all entries in the book. In their totality, they reflect the development of scientific knowledge as emerging "from a nexus of interacting people, agencies, materials, instruments, individual and collective goals/interests, and the histories of all these factors" (McGinn and Roth 1999). This nexus was an intermingling of race, culture, urban poverty, social concerns, academic skills, and environmental design linked by the desire and agency to help transform local conditions. The assessment was a tool for capturing the ongoing creation of this nexus and made visible knowing in science as a process of borrowing and reinvesting knowledge.

Knowledge as Borrowed and Reinvested

With a plan to restore the lot into a usable community space and with seven remaining possibilities (playground, garden, clubhouse, penny store, jungle gym, sandbox, and stage), we drew from the practice of environmental designers in order to advance our practice. What factors do environmental designers consider when planning for an urban space? How can we borrow from that practice? When a local gardening organization presented a slide show to the group, new design considerations emerged, such as a design that required low maintenance, structures that could be built with simple materials (hammers, nails, wood, etc.), and structures that supported social activity for all community members. What could we build that required low maintenance and promoted social exchange for people from age 2 to 102? Design features that were not previously considered became added possibilities, such as a trellis, birdhouse, toolshed, pond, and signs.

While I worked to bring new ideas and resources into the group (the way a classroom teacher might draw upon textbooks and curriculum guides as resources) there was no guarantee that the young people would learn or apply new ideas. To gauge this, I asked the participants to write a response to the question, "What did you learn that might be useful to us now or in the future?" Mia became particularly interested in the concept of signs for representing REAL's message. After the presentation she wrote, "Something that is helpful to us that we can use for the garden is to have signs outside to tell people not to litter, and the mural because it shows what your garden is all about." A month later Mia wanted to develop themes for a mural. (We had received permission to paint a mural on the side of the building that faced the lot.) She came up with several slogans, such as Environmental is fundamental, Help keep the garden clean, Stay in school, Friends forever, and Speak up. The mural themes suggest an ethos of care towards the environment and people. They were messages aimed towards the children living in the community and supported Mia to voice and enact her overall vision for transforming the lot into "a safer place for kids to play."

For Ty, the slides of other community gardens revealed to him "that gardens are in a lot of places" and that "we need a garden house." Garden houses are little replicas of houses that are used by some community gardeners to facilitate social gatherings (some have picnic benches and tables inside). Ty liked the idea of the garden house and later transformed this idea into a "garden shop . . . where we can sell things for gardens like soil, plants, and things." He wanted to move forward and suggested that we needed more support to help clean the lot. He designed a flyer for recruiting volunteers and put his own telephone number on the flyer.

The young people borrowed knowledge and reinvested it in the group, sometimes months later. This level of engagement suggests that we were producing a practice that was both personally and socially relevant to the young people, that was "real." When brainstorming a name for the group, the nature of relevancy became apparent. I asked the young people, What words come to mind when you think about who we are and what we do together? The words generated were designer, activity, community, gardener, service, caring, caring squad, agriculture, helping hands, and environmental. As we played with words, REAL or Realizing Environmental Architecture League was born. Ty did not like "league" and ran downstairs to get a dictionary. He and Mia began reading every word beginning with the letter l. When they came across the word "leader," Ty became very excited and adamant that we change the name. Mia, conversely, wanted to keep "league." To Mia, "league" represented the cooperative and holistic nature of REAL; it fit with her personal mission to change the lives of the children in the neighborhood. Participating in REAL, shoe wrote, "gives me a sense of responsibility and gives me a good feeling about helping people in the community." To Ty, perhaps "leader" represented his sense of his own participation as well as his vision for an entrepreneurial garden. (The name REAL remained, but was later changed to Restoring Environments and Landscapes.) As members of REAL enacted their personal and collective visions, changing identities were quite apparent; an earlier name for the group was "shelter boys."

Assessments often focus on a specific skill or area of knowledge and student's work is judged in isometrics. In this context multiple methods elaborate on single-method findings. Conceptual drawings, measurements, signs, and mural themes were all integral components to the process of environmental design. The resulting products of these activities, however, do not represent the endpoints of learning. Mia's statement of the importance of signs and murals examined in isolation does not illustrate how she revisited and enacted that theme. Their book, then, offers a view of learning as a recursive process where knowledge was borrowed and reinvested into the group and situates knowing in science within personal and social agencies. Further, it not only reflects individual achievements but reflects a cultural shift from "shelter boys" (and girls) to a caring league of environmental architects and leaders. Finally, the book is also a result. The book is a tool that allowed the resources drawn upon to be continuously reinvested into the group, and allowed those resources and the group's history to be public and accessible. In a context where the transience of homelessness meant the group constantly changed, this tool was a valuable asset to maintaining a sense of history and continuance, and to building this practice.

Scientific Plausibility in the Context of Local Action

Gitomer and Duschl describe the "culture" of Project SEPIA (Science Education through Portfolio Instruction and Assessment) as "a learning environment in which students are engaging in learning activities consistent with current psychological, philosophical, historical, and sociological conceptions of the growth of scientific knowledge" (1995, 1). Here, the curriculum reflects what scientists do, including the fluid aspect of their work. "In a portfolio culture, we adopt the view that the goal of science education is to help students develop concepts, theories, strategies, practices, and beliefs that are consistent with scientific ways of knowing, arguing, and exploring" (7). As such, knowledge is seen as subject to change; discoveries are not permanent. Students evaluate existing knowledge, test new ideas, and revise hypotheses. The goal, then, is "to move from the initial diversity of ideas existing in a classroom to a view that represents a consensus by virtue of its scientific plausibility" where the assessment reflects this shift (20). All records become sources of evidence for the assessment and provide rich instructional information. Creating a culture of authentic assessment involves students and teachers reflecting and debating about the standards of good work and the rules of evidence. Students have to demonstrate convincing arguments for their work and what they learned (Lane 1993), communicating their findings to an audience. Maximally, learning continues as debates ensue about factors not considered and new areas of investigation arise (Wolf et al. 1991).

One way to understand the shift toward scientific plausibility is that it represents learning as a process of increased participation in expert practice (Lave and Wenger 1991). This presumes that practice is defined along a linear trajectory of competence from novice to expert (Roth 1998). Levels of student competency are determined based on a hierarchical arrangement of the knowledge and skills that experts utilize. Yet even expert scientists from different traditions offer different explanations to a problem (Linn and Muilenburg 1996). Which discourse and practice of expertise is called upon to make judgments about students' level competency? Using the notion of trajectory of competence ignores the multiple perspectives and experiences that students bring to the solving of problems. It presumes that people are expert or novice and eliminates the possibility that people move in and out of these roles within a given context.

Critical science expands the boundaries of science by challenging who can engage in science, where science can occur, and what constitutes legitimate/expert science knowledge. It brings science to the everyday concerns of people as they engage in transforming human conditions. In the context of critical science, the boundaries of scientific plausibility reflect real-world constraints, such as material resources and time, as well as local values. A basketball court was not plausible because the lot was too small and it did not support activity for all members of the community. The variables that the teens began to envision in 2-D design plans included flower beds, structures that promoted play (chess tables, a stage), and an abundance of garbage cans. As they critiqued each other's designs, they were required to state their reasons for inclusion in the design. Blueberry bushes were plausible "because they look beautiful, can be eaten, and attract birds," and chess tables were equally viable because "old people like to play chess." Some added a path to their design; many liked the path because it added a spatial dimension, separating the play area from the gardening area. In designing an urban environment, the variables that were scientifically plausible were those that had aesthetic, ecological, and social value.

Knowledge as Public and Publicly Enacted

Once a final design was agreed upon, a 3-D model was built. The teenagers worked on various structures, planning the layout of the design and the materials they would need (a process modeled by the landscape designer). They used pictures from magazines as guides. Rocks and twigs were collected from outside to use as the boundaries of the path. Mia sculpted a pond and a birdbath from clay; Ty built flower beds from model wood. The process continued for several sessions; each time people added and/or revised from where others left off. Even youth who rarely participated found a way to contribute (e.g., going to the store for supplies or taking pictures of the model in progress). Quantitatively, the number of structures included in the design plan increased steadily from the initial drawings in October to the conceptual drawings in January to the model in March. Ty's first drawing, for instance, included a laser park and an arcade. In the later 2-D and 3-D designs there were trees, a pond, a stage, flower beds, garbage cans, and more. Qualitatively, the design plan shows how ideas continued to emerge. Through conversations with professionals, visits to local gardens, and pictures of other community gardens, structures such as a trellis, unknown and unimaginable beforehand, became incorporated into the overall design. Further, because products were visual and public, participants (and others) could continuously add to the funds of knowledge that were circulating. For instance, when Ty realized that a chess table was too tall for the scale of the model and unsturdy, he turned it into the stump of a tree by removing the top. Later that day when another participant saw the model she said, "Oh, someone fixed my chess table."

Research examining the use of mathematics and science outside of school suggests that everyday cognition is situated in context, not abstracted from practice, and involves collective efforts of shared reasoning rather than individualized problem-solving attempts (Lave 1988; Resnick 1987; Saxe 1990). However, the competitive nature of schooling often devalues shared reasoning (sometimes called "cheating"), instead supporting individual prowess. As a result, students working in small groups may deny other members access to previously shared knowledge (Kelly and Crawford 1997) and material practices (Roth 1998). In mixed-gender groups, boys spend more time "hogging" science equipment resulting in more passive participation among girls (Jovanic and King 1998). Performance assessments, because they are situated in social interactions and provide multiple opportunities to learn, offer new insights into the shared nature of reasoning, though they cannot by themselves guarantee that equity will be realized (Jovanovic, Solano-Flores, and Shavelson 1994).

The model and other artifacts of REAL were often collectively produced or individually produced toward a collective goal. To the extent that equity was created, it was because these products were public and publicly enacted. That is, young people could add to or borrow from the pool of knowledge that existed. This pool was soon broadened to the larger community, including after-school staff, parents, neighbors, and friends, who also added their feedback. During Community Day the teenagers publicized their design plan using these assessment tools (the book) and the model to represent REAL. In addition to advertising the design plan/model, the goals for that day were to clean the garbage, fix the fence, make signs, and create a fun and productive environment. Ty volunteered to be on the media team to videotape the day as well as interview people. Mia was on the welcoming committee where guests signed the book and were able to have their questions about the project answered. While Mia volunteered for this job she spent most of the day cleaning out the garbage in the lot with a group of teenage girls.

Hours were spent clearing out the debris. Signs were painted with the message, "Help keep our REAL garden clean." The old fence was torn down and posts were made using triangular braces to support the new wire. A group of young people proudly claimed that they "made a pond from natural resources." The trees and existing wildflowers were watered. A DJ set up his equipment so people could work to music. Adults, children, and youth from the shelter worked and played side by side with those from the surrounding neighborhood—a relationship with much historical tension. The refreshment team barbecued hamburgers and chicken and the media team video-interviewed people about the benefits of the garden to the community (a topic they chose). How will this garden benefit the community? Do you think the garden will stay clean? Do you think the fence will get knocked down again? These were some of the realistic questions and concerns that Ty and Josh asked people during the course of day. As they discovered, all were optimistic about the benefits of the garden to the community.

> It's gonna give us sense of responsibility because we're transforming something. We're making something out of nothing. We're gonna be extra proud because we did it. (After-school staff)

> I think it's gonna be a great place for people to play because there's no place right here to just hang out and it's nice to have a stage. I think it's gonna make a difference around here. (Neighbor)

It's gonna turn out to be beautiful. It's gonna help the children take care of the neighborhood by seeing beauty. (Parent)

We have a garden here. Today we did a beautiful job. We're gonna do more but we're gonna take some time doing it. But look around (Ty scans camera around lot)—see how we did. We did a nice job but we need more people to help us. This is (name of after-school program). If you don't live with us, just sign in the black book. You can come help us with the job. Anybody can help us. Anybody can join. (Josh, age fourteen)

All these people have cleaned today. Now it's their time to enjoy! (Ty, age thirteen)

Reform Objectives in Science and Assessment

Was REAL real science? How did the assessment fair in this context? This project draws from the perspectives of science education reform, informal science, environmental action research, children's gardening, critical science, and assessment reform. The objectives of these approaches, listed in table 8.1, are not meant to be exhaustive but represent those often discussed in the literature. There is some overlap, as well as divergence, in the goals of these approaches. In conclusion, I discuss the nature of these reform objectives as they lend themselves to the dialogue on creating inclusive practices in science and offer insights into what that implies about the authentic nature of science and assessment in science, knowing and showing in science, and the purpose of science and assessment.

Authentic Science and Assessment in Science

The American Association for the Advancement of Science envisions an inclusive model for science education, one that expands its boundaries to include mathematics and technology (AAAS 1990, 1993). Informal science also supports this vision though more explicitly seeks to demystify science and technology. Alternative assessments play a role in breaking down the barriers to mathematics and science when tasks are authentic by using everyday objects and reporting students' progress in terms understandable to parents (Kulm and Stuessy 1991). These efforts raise questions about the nature of science and who can participate in it. In the current context, we did not utilize traditional science materials and tools (beakers, test tubes, or microscopes), though we might have if it emerged as interesting and necessary to pursue in the course of reaching our objectives. We did engage in science as part of everyday urban life and

Table 8.1 Objectives of Science from Several Perspectives

Approach/Model	Objectives
Science education reform (AAAS 1990, 1993)	Familiar with the natural world and respecting its unity; aware of ways in which math, technology, and science interact; have a capacity for scientific ways of thinking; know that science is a human enterprise; use scientific knowledge and ways of thinking for personal and social purposes.
Informal science (Crane 1994; Hoftstein, Bybee, and Legro 1997)	Renew interest in science; demystify science and technology; provide opportunities for underrepresented groups to explore science.
Environmental Action Research (Ahlgren and Rutherford 1993; Eisenhart, Finkel, and Marion 1996)	Create a socially responsible science; develop questioning attitude; develop sense of active agency to participate in science toward sociopolitical ends; use scientific methodology.
Children's gardening (Kiefer and Kemple 1998)	Develop appreciation of natural world; understand interdependency of nature; develop leadership and teamwork skills; promote community interaction; support neighborhood beautification; change perceptions toward urban youth.
Critical science (Barton 1998b; Rodriguez 1998)	Be inclusive of multiple perspectives and ways of knowing; expand boundaries of science; create reflexive relationship between individual and science.
Assessment reform (Baron 1991; Gitomer and Duschl 1995; Kulm and Stuessy 1991)	Provide equitable opportunities to demonstrate learning; assess reasoning in context; use authentic tasks; close gap between assessment and instruction; assess higher-order skills; reflect and debate on the standards of good work; support sustained inquiry and constructivist pedagogy.

community by using available land (an empty lot) and access to greening organizations that supplied plants, wood, building materials and equipment, and technical assistance. Technology and mathematics were not supplanted by design and gardening; they enhanced production. Informal and formal mathematical strategies were utilized to measure, design, form symmetry, build, and plant. Technology offered a method for creating flyers, signs, reports, and letters, and media equipment facilitated the documentation of our collective history.

The young people participated in a variety of tasks that were needed in order to get the work done. Drawing, for instance, was an activity incorporated into their performance as designers because it is what designers do. While drawings are increasingly being used in the upper grades to measure science learning (Belynne-Buvia 1996; Fellows 1994), in this context, drawing was not introduced to measure outcomes or assess their understanding of science concepts; it was an inseparable component of the program. The various performances were the assessments and thus were also authentic because they were inseparable from the co-production of science.

Further, to the extent that the book was an authentic assessment it was due to participation of young people in its creation and its public nature. As Gipps recently states, "The best defense against inequitable assessment is openness. Openness about design, constructs, and scoring will bring out into the open the values and biases of the test design process, offer an opportunity for debate about cultural and social influences, and open up the relationship between the assessor and learner. These developments are possible, but they require political will" (1999, 385). The performances represented in the book were all-inclusive and could be viewed and enacted publicly. The young people used these tools to represent their achievements, e.g., putting the model and the book on display during Community Day and other events at the shelter. The book, then, circulated and interacted with the community. Parents could readily see the work produced by their children as they participated in designing and implementing environmental change, and their own voices were included in the book as they provided feedback during events such as Community Day. Of course, this openness requires that the adult facilitator include him/herself in the possibility of critique. It requires that the adult view him/herself as included in the context within which opportunities for learning were created. That is, it requires a view of learning and assessment as relational activity.

Knowing and Showing in Science

As Shepard (1991) points out, there is a clear division within the measurement community as to the types of tests that should be constructed in relation to the types of learning that are valued. Grounded in behavioral learning theory, content-driven or criterion-referenced tests assume that basic tenets of knowledge should be mastered before higher-order skills can be introduced. Learning, from this vantage point, is seen as linear and sequential. Conversely, authentic assessments are constructed in a manner consistent with current (and currently popular) understandings of learning and with what it means to "use a mind well." That is, from a constructivist framework, learners participate in knowledge-building communities where meaning is constructed, not imparted; discussion of new ideas and multiple modes of expression are fostered; and the culture of expert practice is created and maintained through collaboration (Collins, Brown, and Newman 1989; Rogoff 1990). Content is brought to life in the context of real-world applications.

In science education, practices are developing that have exposed the need to rethink the unit of analysis. Science classrooms are being transformed into scientific communities where students collectively muddle in ill-defined problem frames, such as engineering (Roth 1998), technology (McShane and Yager 1996), and ecological research (Eisenhart, Finkel, and Marion 1996), to name a few. Scientific knowledge is seen as emerging from a nexus of people as they interact with/in society and history. This account of learning renders individualistic cognitive accounts as insufficient for understanding the dynamic, recursive process of learning as socially and culturally based (Rogoff and Chavajay 1995). In such environments, individual contributions toward a collective product are difficult to isolate. The implication for assessment is that work collectively produced should be understood in relation to what it is and how the group produced it.

The method that emerged here toward this end was a historical account of production. The documentation of REAL's history shows how knowledge emerged and was reinvested into the collaborative practice. New ideas did not replace old ideas but offered the occasion for old ideas to be reinvested in new ways. For instance, during the slide show one youth learned that storage was important because "we need a place for our tools"; in the construction of the model, he designed a storage space underneath the steps to the stage. Signs were introduced as a way to advertise the name of the group and hours of operation; later, the mural was considered a sign for telling people about the theme of the garden. Viewing the design over time, quantitative and qualitative changes are evidenced. Quantitatively, the number of structures included in the design plan increased steadily. Qualitatively, the design illustrates a transformation from the modern entertainment of cyberspace games and sporting events to community activities that promote interaction and add beauty to the environment. The book illustrates these gradual changes over time in a practice that borrowed from landscape designers, gardeners, journalists, etc. Knowing in science was situated in a cultural and historical account of production where individual contributions were interwoven in this complex history and social nexus.

The Purpose of Science and Assessment

Science education reform expands the boundaries of science by including science that is a human enterprise and can be used for personal and social purposes. While science education reform has centered its mission around the creation of a science for all, many critical science educators have argued that if science is to be for all Americans, methods for connecting to and including a variety of children's experiences will be necessary. Environmental action research, children's gardening, and critical science add a political dimension to reform by using science and scientific tools to transform human conditions. In environmental action research, students investigate the local environment and develop action plans for changing policy and/or practice. The premise of socially responsible science education lies in the theoretical and practical work of critical and radical pedagogy. Writers within these traditions argue that the ideological biases of education must be uncovered as part of the political struggle against oppression of Euro-American domination (hooks 1994; Shor 1992). The goal becomes to help students think critically about issues related to science and question the appropriateness of scientific methodologies within the context of political or economic self-interest (Ahlgren and Rutherford 1993; Eisenhart, Finkel, and Marion 1996).

The purpose of REAL was to transform the physical and social landscape, to engage in science and research as part of urban life and community. Revitalizing of the natural world and understanding the interdependency of life (physical and social) offered a view of science as a means for improving human conditions, in addition to "the contemporary school view that society requires technological innovations based on rational models of science to advance" (Kiefer and Kemple 1998, 106). From a critical science perspective, the young people were producers of a science that meant something to them. Science was not an abstract body of knowledge to be learned but was "something to be proud of, to be remembered by, and to help beautify the community," as one youth put it.

In her book *The Having of Wonderful Ideas*, Duckworth asks, "What happens to children's curiosity and resourcefulness later in their childhood? Why do so few continue to have their own wonderful ideas?" (1987, 6). She suspects, as many others have, that the answer lies in the nature of schooling where students' ideas come to be less and less valued. In school, students are required to reproduce the knowledge espoused by teachers, textbooks, and tests. They are related to as having little to give to their own educational process; adults produce the environment *wherein* children learn and develop. Homeless children in particular are vulnerable to silencing and anti-democratic practices (Quint 1994). Science remains an abstract body of knowledge in a world distinct from their own. Conversely, valuing children's experiences "shifts the dynamics of what counts as science and who can do science because children would not have to silence certain experiences or feelings traditionally labeled outside of science" (Barton 1998b, 386).

The project offered an occasion for students' wonderful ideas to be the impetus of reflection and action. Expressions and ideas were the impetus to building a community-based science. One boy distinguished this from school when he said, "I thought it was gonna be like a project, like in school, you know like a fake project." The conception of "a fake project" raises the question, Toward what end are students doing science in school? Here, science was supported by a non-Western vision; it was socially oriented rather than task oriented (McShane and Yager 1996). Science was not defined in relation to the perspectives of urban youth living in poverty; it was redefined as the young people worked to transform a community, "even though they won't live here for a long time," as Ty put it. Curiosity and resourcefulness emerged from being connected/reconnected to themselves as creators of a science in the service of community.

How could the assessment also be real, or produced for a particular social and cultural purpose that was relevant to the lived experiences of the young people? Creating a performance assessment within the context of critical science required a critical reflection of the purpose of assessment that emerged alongside the development of practice. Was the purpose of assessment to examine students' understanding of science? If so, whose science and what would count as science understanding? In the context of local action, toward what end does the assessment serve? Documenting the process by which we progressed as a group, the things we accomplished and learned along the way, and the performances which young people developed was a method that was constructed in order to document practice *and* build an ongoing process where all young people could participate. The flexibility of the program, as well as the transience of homeless youth, meant that participants appeared, disappeared, and reappeared. They were not systematically involved in the same process. As teenagers and children arrived at the shelter and/or entered the program, the book served as a means for publicly illustrating the history of the project. Young people of all ages and skill levels could join in the performance as designers using what they had to give. Some drew; others took photographs. As Josh stated, "Anybody can help us. Anybody can join."

REAL's history was not about representing homeless youth's understanding of science. It was about creating something new with them. The documenting of REAL's history was one of the many activities we created together. As the young people created community and a community garden, their collective history was documented through photographs, video, writings, drawings, interviews, etc. Methods, such as drawings, emerged not because they were more accurate measures of science understanding but because they were part of the improvisational performance as designers. Viewing assessment reform from this perspective necessitates a descriptive, rather than explanatory, analysis of change. Here, there is no blueprint for learning and assessment. Teacher actions in one specific situation cannot be likened to teacher actions in another to produce similar results (McNiff 1993). Creating an inclusive practice may mean developing a performance assessment that involves a month-long study of urban pollution, or it may mean testing chemical reactions through step-by-step procedures. Here, an inclusive science practice and assessment were created from the particular interests, ideas, and lived experiences of the young people. The assessment was a tool for building our practice as well as the result of real-world community-based science. Its value to the group emerged and reemerged in different and unpredictable forms. The assessment often served as a catalyst for dialogue about the project. It came with us on trips, on the subway; it was on display during community events; it was a mechanism for getting feedback and suggestions from participants, parents, and other community members. It was a method that allowed the young people to represent what they had created and allowed new members to join in creating its future. In essence, they were the founders and producers of a community of science/science for community practice, simultaneously making and writing their history.

References

Ahlgren, A., and Rutherford, J. 1993. Where is Project 2061 today? *Educational Leadership* 50: 19-22.

American Association for the Advancement of Science (AAAS). 1990. *Science for all Americans.* New York: Oxford University Press.

———. 1993. *Benchmarks for science literacy.* New York: Oxford University Press.

Baker, E. L. 1994. Researchers and assessment policy development: A cautionary tale. *American Journal of Education,* 102: 450-477.

Baker, E. L., O'Neil, H., and Linn, R. L. 1993. Policy and validity prospects for performance-based assessment. *American Psychologist,* 47: 1210-1218.

Baron, B. J. 1991. Performance assessment: Blurring the edges of assessment, curriculum, and instruction. In G. Kulm and S. M. Malcom (Eds.), *Science assessment in the service of reform* (pp. 249-265). Washington, DC: American Association for the Advancement of Science.

Barton, A. C. 1998a. *Feminist science education.* New York: Teachers College Press.

———. 1998b. Teaching science with homeless children: Pedagogy, representation, and identity. *Journal of Research in Science Teaching,* 35: 379-394.

Belynne-Buvia, K. 1996. The mission to Mars drawing lesson: A schools for thought pilot study. Paper presented at the annual meeting of the American Educational Research Association, April, New York, NY.

Bogdan, R. C., and Biklin, S. K. 1992. *Qualitative research for education: An introduction to theory and methods* (2nd ed.). Boston: Allyn and Bacon.

Collins, A., Brown, J. S., and Newman, D. 1989. Cognitive apprenticeship: Teaching the crafts of reading, writing, and mathematics. In L. B. Resnick (Ed.), *Knowing, learning, and instruction: Essays in honor of Robert Glaser* (pp. 453-494). Hillsdale, NJ: Erlbaum.

Crane, V. 1994. An introduction to informal science learning and research. In V. Crane, H. Nicholson, S. Bitgood, and M. Chen (Eds.) *Informal science learning.* Dedham, MA: Research Communications, Ltd.

Duckworth, E. 1987. *"The having of wonderful ideas" and other essays on teaching and learning.* New York: Teachers College Press.

Eisenhart, M., Finkel, E., and Marion, S. F. 1996. Creating the conditions for scientific literacy: A re-examination. *American Educational Research Journal,* 33: 261–295.

Fellows, N. J. 1994. Into the woods: Nature experience and conceptual change. Paper presented at the annual meeting of the American Educational Research Association, April, New Orleans, LA.

Gipps, C. 1999. Socio-cultural aspects of assessment. In A. Iran-Nejad and P. D. Pearson (Ed.). *Review of research in education,* Vol. 24 (pp. 355–392). Washington, DC: American Educational Research Association.

Gitomer, D. H., and Duschl, R. A. 1995. Moving toward a portfolio culture in science education. Technical Report for the Center for Performance Assessment. Princeton, NJ: Educational Testing Service.

Greeno, F. Reif, A. Schoenfeld, A. Disessa, and E. Stage, 219–34. Hillsdale, NJ: Lawrence Erlbaum Associates, Publishers.

Herman, J. L., Aschbacher, P. R., and Winters, L. 1992. *A practical guide to alternative assessment.* Alexandria, VA: Association for Supervision and Curriculum Development.

Hoftstein, A., Bybee, R., and Legro, P. 1997. Linking formal and informal science education through science education standards. *Science Education International,* 8: 31–37.

Holzman, L. 1997. *Schools for growth: Radical alternatives to current educational models.* Mahwah, NJ: Lawrence Erlbaum Associates, Publishers.

hooks, b. 1994. Transformative pedagogy and multiculturalism. In T. Perry and J. Fraser J. (Ed.) *Freedom's plow: Teaching in the multicultural classroom* (pp. 91–97). New York: Routledge.

Jovanovic, J., and King, S. S. 1998. Boys and girls in the performance-based science classroom: Who's doing the performing? *American Educational Research Journal,* 35: 477–96.

Jovanovic, J., Solano-Flores, G., and Shavelson, R. J. 1994. Performance-based assessments: Will gender differences in science achievement be eliminated? *Education and Urban Society,* 26: 352–366.

Kelly, G. J., and Crawford, T. 1997. An ethnographic investigation of the discourse processes of school science. *Science Education,* 81: 533–559.

Kiefer, J., and Kemple, M. 1998. *Digging deeper: Integrating youth gardens into schools and communities.* Vermont: Common Roots Press. A comprehensive guide produced by Food Works in partnership with the American Community Gardening Association.

Kulm, G., and Stuessy, C. 1991. Assessment in science and mathematics education reform. In G. Kulm and S. M. Malcom (Eds.) *Science assessment in the service of reform* (pp. 71–88). Washington, DC: American Association for the Advancement of Science.

Ladson-Billings, G. 1994. *Dreamkeepers: Successful teachers of African American children.* San Francisco: Jossey-Bass.

Lane, S. 1993. The conceptual framework for the development of a mathematics performance assessment instrument. *Educational Measurement: Issues and Practice,* 12: 16–23.

Lave, J. 1988. *Cognition in practice: Mind, mathematics, and culture in everyday life.* Cambridge, UK: Cambridge University Press.

Lave, J., and Wenger, E. 1991. *Situated learning: Legitimate peripheral participation.* Cambridge, UK: Cambridge University Press.

Linn, M. C. and Muilenburg, L. 1996. Creating lifelong science learners: What models form a firm foundation? *Educational Researcher,* 25: 18–24.

Malcom, S. M. 1991. Equity and excellence through authentic science assessment. In G. Kulm and S. M. Malcom (Eds.) *Science assessment in the service of reform* (pp. 313–330). Washington, DC: American Association for the Advancement of Science.

McGinn, M. K., and Roth, W. M. 1999. Preparing students for competent scientific practice: Implications of recent research in science and technology studies. *Educational Researcher,* 28: 14–24.

McNiff, J. 1993. *Teaching as learning: An action research approach.* New York: Routledge.

McShane, J. B., and Yager, R. E. 1996. Advantages of STS for minority students. In R. E. Yager (Ed.) *Science/technology/society as reform in science education* (pp. 131–138). New York: State University of New York Press.

Nickerson, R. S. 1989. New directions in educational assessment. *Educational Researcher,* 18: 3–7.

Noffke, S. E. 1997. Professional, personal, and political dimensions of action research. In M. W. Apple (Ed.) *Review of research in education*, Vol. 22 (pp. 305–343). Washington, DC: American Educational Research Association.

Quint, S. 1994. *Schooling homeless children: A working model for America's public schools.* New York: Teachers College Press.

Resnick, L. 1987. Learning in and out of school. *Educational Researcher,* 16: 13–20.

Rodriguez, A. J. 1997. The dangerous discourse of invisibility: A critique of the National Research Council's National Science Education Standards. *Journal of Research in Science Teaching,* 34: 19–37.

———. 1998. Strategies for counterresistance: Toward sociotransformative constructivism and learning to teach science for diversity and for understanding. *Journal of Research in Science Teaching,* 35: 589–622.

Rogoff, B. 1990. *Apprenticeship in thinking.* New York: Cambridge University Press.

Rogoff, B., and Chavajay, P. 1995. What's become of research on the cultural basis of cognitive development? *American Psychologist,* 50: 859–877.

Roth, W. M. 1998. *Designing communities.* Boston, MA: Kluwer Academic Publishers.

Roth, W. M., and McGinn, M. K. 1998. UnDELETE science education: Lives/work/voices. *Journal of Research in Science Teaching,* 35: 399–421.

Ruiz-Primo, M. A., and Shavelson, R. J. 1996. Rhetoric and reality in science performance assessments: An update. *Journal of Research in Science Teaching,* 33: 1045–1063.

Saxe, G. B. 1990. The interplay between children's learning in school and out-of-school contexts. In *Toward a scientific practice of science education,* ed. M. Gardner (Ed.). Hillsdale, NJ: L. Erlbaum.

Shepard, L. A. 1991. Psychometricians' beliefs about learning. *Educational Researcher* 20: 2–16.

Shor, I. 1992. *Empowering education: Critical teaching for social change.* Chicago: University of Chicago Press.

Vygotsky, L. S. 1978. *Mind in society.* Cambridge, MA: Harvard University Press.

Wolf, D., J. Bixby, J. Glenn III, and H. Gardner 1991. To use their minds well: Investigating new forms of student assessment. *Review of Research in Education* 17: 31–74.

Note

1   The project was funded by the National Science Foundation (#9733700) under the principal investigation of Angela Calabrese-Barton. Courtney Desmond St. Prix was a co-producer of the teen project.

Chapter 9

Guinea Pig Pedagogy: Critiquing and Re-embodying Science/Education from Other Standpoints

Matthew Weinstein

This chapter argues for the importance of multiple, positioned narratives in science education. More specifically, it argues that science teachers and science education researchers should take into account the perspective of human research subjects (and research subjects more generally) within their pedagogical and scholarly works. The subject location of the object of study is clearly centrally important to scientific work, yet it has been rendered marginal and more often invisible in the enterprise of science education writ large.

This move of discursively erasing what is materially central is not unique, of course. The labor of white women and African slaves provided a substantial base for nationhood and "democratic republican" ideals in American society—a labor extractable in part only through control of discourse, i.e., through isolation and censuring and censoring technologies. While it would be incorrect to say that the language, art, and other expressive media of such groups was entirely silenced, it would not be wrong to say that they were hushed, trapped, and unheard; this is the meaning of marginalization in my story. Such marginalization permitted vulnerability and limited the means of these groups to speak back or have their exploitation witnessed by those who might be allies.

Scientific cultures have counted on this marginalization. The marginal, whoever that might be at any historical moment, have consistently formed the source for human research subjects, e.g., for pharmaceutical, medical, and psychological experimentation. At various times this has included orphans, widows, both urban and rural African Americans, conscientious objectors, prisoners, and the poor more generally.[1]

As with marginalized populations generally, human subjects in the 1960s and 1970s began demanding rights. These struggles were part and parcel of civil rights movements of women and racial/ethnic minorities. In the wake of these struggles, in the 1980s and 1990s there emerged a full-blown public culture of human subjects or, as some more political members of that "community" have preferred to be called, guinea pigs.[2] Web pages, zines (independently produced and usually anticorporate magazines), and other public forms of networking now exist for guinea pigs.

After a brief discussion of the problem of extant standpoints in science education, this chapter weaves together a brief overview of guinea pig history and politics with a discussion of feminist standpoint theory, an approach to knowledge which helps theorize the possible role and importance of guinea pig (and other marginalized) points of view in the science curriculum. The two threads, standpoint theory and guinea pig theory, are presented in a woven pattern rather than sequentially since neither perspective has an a priori status. However, even though guinea pig politics has not been substantially informed by standpoint theory, nor have standpoint theorists seriously analyzed guinea pig history, the two accounts of knowledge support and substantiate each other in playful ways. This overlaying of multiple, distinct, interacting accounts is referred to by Donna Haraway as diffraction, a concept developed in detail later in this chapter. I follow this double braid with yet another example of diffraction which makes my point (i.e., the value of multiple, contesting accounts) even more explicitly: a triple braid of narratives about digestion—digestion as narrated by the science standards, medical history, and guinea pig politics.

Natural Knowledge in Schools:
A Problem of Perspective

My argument about the need for multiple, differently located (in the social sense of the word) perspecitves presumes that reformed science education currently lacks such multiplicity. This multiplicity is critical if science education is to be meaningful and empowering to a broad and, in the view of many, increasingly heterogeneous body of students. By reformed science education I mean science education that seems to live by the letter and spirit of current reform movements, especially as written in the

National Science Education Standards recently proposed by the National Research Council, a document that has largely been accepted as the benchmark of high quality science education by most state and academic agencies. The *Standards* are the end result of ten years of struggle by professional groups to reach a consensus on what science education should look like.[3] It is, as a result, a complex, often contradictory, "heteroglossic" text. I will not provide here a thorough analysis of the *Standards*, what it offers and lacks. My purpose here is much narrower. I wish only to comment on a general standpoint issue within the text of the *Standards*.

The general knowledge standpoint, which for the moment might be taken as synonymous with point of view, of the *Standards* is stated baldly in the document's section titled "Principles and Definitions." "School science reflects the intellectual and cultural traditions that characterize the practice of contemporary science". In theory this phrase "practice of contemporary science" could mean many things; it could include all the different roles and activities involved in making scientific knowledge, including those of publishers and janitors, spouses and secretaries, bosses and technicians, as well as scientists themselves. This picture of a widely spread network involving everyone from senators to consumers, who together provide the necessary material conditions for scientific work, is supported by recent sociology of science. Instead, that word "practice" clearly refers only to the work of scientists. No one else is ever mentioned in the *Standards*. The point is hammered home through the photos that fill the *Standards* and feature students, mostly minorities, dressed up (more or less) as scientists "inquiring." These photos act as a sort of a reversal of Roland Barthes dictum that caption text serves to "fix the floating chain of signifieds in such a way as to counter the terror of uncertain signs", i.e., the ambiguous meanings of any photo. Here the photos fix the meanings of the text, illustrating what characterizes the practice of contemporary science: lab work by people wearing white coats.

The image of science reflected in the photographs is doctored in two ways. First, as already noted, the reflection excludes all the other people that make lab work possible: janitors, guinea pigs, spouses, secretaries, and funders. These groups also have traditions that are part of, not separate from, those of science but that are rendered invisible here. Second, the image is cleansed of the messy historical and human relations that have been part of scientific history: racism, sexism, and more general forms of exploitation (culling of guinea pigs from vulnerable populations, colonialism, etc.) as well as more innocuous cultural and historical contingencies. These are simply excluded a priori. The reformed science education can thus be seen as doubly selective : first, erecting a boundary around scientists so that they alone are inside, doing science; second, cleaning that interior arena so that science is a completely cerebral and methodological activity "characterized by empirical criteria, logical argument, and skeptical review".

This vision of science, which is unsupported by sociological and anthropological studies of scientific work, purifies scientific knowledge of its historical and social entanglements. Cut free, a romantic vision of scientific knowledge and method risks becoming unitary, uncontestable truth. How do we re-tether science so that it can be understood in its context? This has been the project of a group of scholars working out the concept of a knowledge standpoint.

Standpoint Theories

Standpoint theories offer a direct contradiction and confrontation to versions of science which imagine knowledge as independent of the knower, of history, or of culture. Yet there are multiple versions of what constitutes a knowledge standpoint, and these involve conflicting ideas about objectivity, identity, and politics. Here I will trace out what must be a personal and somewhat idiosyncratic path through these debates by looking at ideas about standpoint knowledge as elaborated by Sandra Harding, Donna Haraway, and Maria Lugones.

Standpoint theories emerged in multiple fields more or less independently in the late 1970s and early 1980s. From that period, Dorothy Smiths' version is one of the most often referenced and is one of the more sophisticated variants. Smith's project was (and continues to be) the reconstruction of sociology around the interests of women. Drawing on a feminist revision of Marxist materialism, she has argued that sociological knowledge comes from problematizing the quotidian of women's lives in all of its material specificity: Look at real existing situations and explore how they are established, what relations are rendered invisible, and who is served by such invisibility. The value of such a reconstruction is that oppressive structures become visible through this analysis, structures which are invisible in mainstream sociology, invested as it is in current, patriarchal power relations.

For Smith science is merely one more institution at stake. Only once in *The Everyday World as Problematic* does science come up by name: "[Previously] I explored issues for women arising from a culture and politics developed almost exclusively by men and written from the standpoint of men and not of women. This statement was as true of intellectual and *scientific* discourses as of TV commercials" (107, emphasis mine). It is philosopher Sandra Harding who brings standpoint theory to bear on scientific knowledge production itself.

Harding reverses the usual assumptions about neutrality and objectivity as they are enunciated in scientific mythology, i.e., that detached, unpolitical science is objective science. Because status quo science, which is supposedly committed to objectivity, refuses to see, by intent or mere circumstance, structures of oppression, objectification, and exclusion, Harding marks it as less objective. Conversely, because feminist and other standpoint knowledges are able to observe and track human suffering and its structural causes, they are posited as more objective. In fact, Harding constructs what amounts to an objectivity continuum around this reversal, with weak objectivity (or objectivism) at one end, i.e., the commonsense notion of objectivity, and strong objectivity at the other end, i.e., an objectivity which adds to weak objectivity the multiple standpoints of "Others" excluded from the practice of science. "To enact or operationalize the directive of strong objectivity is to value the Other's perspective and to pass over in thought into the social condition that creates it—not in order to stay there, to 'go native' or merge the self with the Other, but in order to look back at the self in all its cultural particularity from a more distant, critical, objectifying location". The moral is clear: for the most objective science, go out and solicit the knowledges of Others, not individual by individual, but, Harding makes clear, as it is produced in collective social struggle.

In the previous quotation it is clear that the perspective and the standpoint ("the social condition that creates it") are not identical. And yet there is a slipperiness in Harding's philosophy, so that elsewhere this process is described as starting "thought in the perspective from the life of the Other, allowing the Other to gaze back "shamelessly" at the self who had reserved for himself the right to gaze "anonymously" at whomsoever he chooses" (1991, 150). Here, through the identification of standpoint with perspective and gaze, the distinction is lost. Standpoint is entirely defined by the sort of visual metaphors from which it is elsewhere distinguished.

More problematic, however, is defining whose voices and whose struggles define a standpoint and how those not sharing particular standpoints can come to learn from them. In theory, strong objectivity is available for everyone. " [S]tarting thought from women's lives is something that both men and women must learn to do. Women's telling their experiences is not the same thing as thinking from the perspective of women's lives". However, it is not clear at all which standpoints in which order—from the million of standpoints available, i.e., theorizing of women's' lives and other Others' lives—to include, nor is it likely that combining such theories would result in anything approaching the unity and coherence that a term like "strong objectivity" implies, i.e., supposedly less "distorted" (143). As Longino notes in her review essay of Harding and other standpoint theorists, "[I]t is not clear to me that there is any neat continuum. I cannot produce thought from the life situations of women in India, although I can produce thought that takes their point of view and research as seriously as theirs takes my own. However much I and they inform ourselves about one another's life situations we can neither share nor escape our social locations unless we materially dismantle them, and even then we cannot escape our histories. There is, therefore, no guarantee of a convergence of theory, even when it meets Harding's criterion for maximal objectivity". Part of the problem is that knowledge in this perspective seems reified as something combinable. Each of the summands in the equation of strong objectivity are the knowledge products of the struggles of identity politics. All those Others, including the feminist Other, seem discrete, impermeable, and thus ahistorical. Some other way of framing this knowledge is needed if the desired goal of a socially just science is to be reached.

Guinea Pig History (*Guinea Pig Zero (GPZ)*)

One way of reading guinea pig history in the United States is to read it backwards and forwards from the Belmont Report of 1979.[4] It is this report that established the current system of informed consent for people participating in research, institutional review boards (IRB's), and participation in research as a voluntary activity. I call the current system of post-Belmont human subject work "modern guinea pigging." The Belmont Report was the system's response to a number of scandals that came to light in the 1970s. These included the exposure of the Tuskegee Experiment in the early 1970s in which 399 African American men were allowed to suffer through advanced stages of syphilis merely to see the effect that spirochetes would have on black men (the effects on white men were established at the turn of the century) and the Jessop, Maryland, prisoners' law suit, in which it was revealed that they had been exposed to viral diarrhea, shigella, typhoid, cholera, and malaria in exchange for better accommodations and two to four dollars a day. These scandals and actions forced the United States to abide, as other countries already had, by a code of ethics at least rising to the standards of the Nuremberg Code established at the end of World War II.

After the Belmont Report, guinea pigging became more of an economic affair. Research institutes ("units" as they are called by the guinea pigs) would typically have a paid recruiter responsible for advertising for subjects and subsequently screening them. These recruiters would also inform guinea pigs of the possible benefits and risks involved in experimental procedures.

Of course this is an idealization. A lot of research proceeds without consent or concern for subject safety in the United States: "In a shocking [1995] study, a White House panel ask[ed] 1800 randomly selected patients in hospital and clinic waiting rooms around the country whether or not they were participating in medical research. 412 of these patients turned out to be research subjects, including 99 who said they were not and never had been. In other words, close to 25% of the people surveyed who were participating in medical research did not realize it . . . About 20% of these unknowing participants were involved in experiments that posed more than a 'minimal' health risk".

Scandals continue to erupt. Psychiatric patients have been forced to be guinea pigs in a Veteran Administration hospital in Augusta, Georgia. Scottish student guinea pigs were exposed to organophosphate pesticides known to have low-level effects. In 1989, Hispanic and African American children were immunized with the EZ measles vaccine in Los Angeles, and their parents were not informed that the vaccine was not licensed in the United States. This added to the alienation between the medical establishment and communities of color still reeling from the Tuskegee experiment. The government organizations involved in monitoring the treatment of subjects are only beginning to hold research facilities accountable, as came to light with the May 1999 closing of Duke University's research facility by the Office of Protection from Research Risks.

With the conversion of guinea pigging from an involuntary, unregulated activity to one involving recruitment and economic motivation, guinea pigging has become a job. While for most guinea pigs it is at most an incidental form of income, for others it is a major form of support.[5] A stable core of guinea pigs has even formed in research dense areas. As Robert Helms, a guinea pig activist, notes, recruiters desire this stable, professional guinea pig population: "Try to imagine how much time it would take to fill [beds] with all brand new people who had never done it [experiments] before because you've got to explain, you've got to take their histories, you don't have a file on them . . . and you have to explain how it works. You have to watch half of them chicken out, 'cause they don't know what they're getting into, and half of them are going to pass out . . . They have jobs to do. My blood is just as good as an inexperienced guy's blood, and I'm not going to go and eat a damn poppyseed bagel and get bounced from the study . . . So it's better to have regular guinea pigs".

As a result of its restructuring as work, being a human subject has become vulnerable to shifts in the medical industry, particularly consolidation. Where there used to be many smaller research labs, there are now only a few larger units. Simultaneously, guinea pigging has gone from an obscure job to one that is known. Helms captures some of the effects that this dual trend has had on working conditions: "They're also making bigger units, and there's over crowding. They're packing the studies one after another, really tight. So sometimes we'll be sleeping on cots, folding cots, except during the times when we're actually dosing, and I got paid extra 25 bucks a day to sleep on an easy chair, a recliner in one place. And the place was mobbed, you couldn't get any quiet no matter what time it was. Even in the middle of the night it was noisy as hell. It was more than twice the number of people that the place was made for. You couldn't get a hot shower, cause there's too many people showering". Such conditions make guinea pigging ripe for organized occupational politics. This has been the extracurricular work of Helms, who has published a magazine, *Guinea Pig Zero*, on the history, politics, and culture of guinea pig life.

Situated Knowledge

Harding's strong objectivity is simple, political, and highly problematic. Its simplicity and politics provide a clear directive to those seeking a more socially just science: join in struggles for democratic participation and include the stances of Others to increase the objectivity of your science. As noted, such inclusion is undertheorized in Harding. It is also important that the two portions of strong objectivity (political struggle, inclusion of others' voices) are not mutually dependent: one can include others' voices without engaging in political work, and visa versa.

Donna Haraway engages each of these in her review of Harding, titled "Situated Knowledges". In this article, Haraway lays out a radical revision to Harding's standpoint theory. Inspiring these revisions is a greater intimacy with and admiration of scientific practice (Haraway's background is in theoretical biology) and a philosophic pragmatism refracted from the work of Ian Hacking. This pragmatism is played out in part in a postmodern mix of theories whose contradictions are self-consciously left standing. This mix includes social studies of scientific knowledge, feminist and Marxist standpoint theory, scientific realism, and other feminist and antiracist cultural analyses. Against Harding's continuum of objectivities, with a holistic, inclusive goal as the desired aim, Haraway argues for the partiality of all knowledge claims. The real danger for her is knowledge that refuses accountability by not clarifying the naturo-techno-social processes that produced it—and for Haraway all knowledge is produced. "The moral is simple: only partial perspective promises objective vision". This is not, however, a call for relativism; there are forbidden positions here: those that mask themselves as totalities, a rhetoric that Haraway calls "the god trick" (189). I read Haraway in this piece as positioning strong objectivity as one example of such a totality. "Not just any partial perspective will do; we must be hostile to easy relativisms and holisms built out of summing and subsuming parts" (p. 192). Rather than seeking out the usual suspects of gender, class and race, Haraway argues that we must "seek perspective from those points of view, *which can never be known in advance*, which promise something quite extraordinary, that is, knowledge potent for constructing worlds less organized by axes of domination" (192, emphasis mine). These points of view may of course be those valued by traditional standpoint theory (191). But Haraway says, they may also emerge from Nobel award winning immunology, cyberpunk science fiction, or in genetically engineered mice. To distinguish this from the model Harding presents Haraway calls hers a theory of situated knowledge. The heart of the difference emerges in Harawy's cautionary note to standpoint theorists: "[F]eminist embodiment resists fixation and is insatiably curious about webs of differential position. There is no single feminist standpoint because our maps require too many dimensions for that metaphor to ground our visions". Here the dichotomy that haunts Harding (and Smith) between standpoint as a visual metaphor and standpoint as a theory emerging out of multiple specificities is subverted. Haraway in one swoop rejects a stereotyped woman's or feminist position while embracing the visual metaphor.

At the same time, Harding's goal of political struggle over science remains in focus. "The feminist standpoint theorists' goal of an epistemology and politics of engaged, accountable positioning remains eminently potent". Coalitions and democratic action are the bottom line in both accounts of knowledge. In recent years Haraway has clarified her connection and differences with Harding. In fact there is a clear rapprochement, with both theorists citing the other and Haraway explicitly using the language of "strong objectivity" and "standpoint theory" in conjunction with situated knowledge.

Haraway succeeds in de-reifying the standpoints of Smith's and Harding's theory. Situated knowledge is mobile, contingent, and does not claim truth value merely because of a metaphor of location, but rather through the systems of accountability it brings into play, the ways it makes itself partial in both senses (limited and political) of the word. Yet, how we see from these multiple positions still remains largely mysterious here. It is in others' works that such technologies are developed.

Guinea Pig Politics

Another way of reading guinea pig history draws attention to the formation of a guinea pig public culture in 1995 with the publication of *Guinea Pig Zero* by guinea pig and anarchist activist Robert Helms. Anthropologists use the term public culture to refer to public displays which self-consciously try to present, represent, and define culture; public culture includes such media as museums, magazines, television, films, parades, restaurants, and curricula. As noted, *Guinea Pig Zero* (*GPZ*) is a "zine:" an informally produced, anticorporate magazine. According to *The Book of Zines*, zines are "Cut-and-paste, 'sorry this is late,' self-published magazines reproduced at Kinko's or on the sly at work and distributed through mail order and word of mouth. They touch on sex, music, politics, television, movies, work, food, whatever". And "Factsheet 5," a well recognized review of zines, defines them as "small press publications with a run of 15–2,500. They often deal with obscure or controversial subjects, or they're about the life of the publisher, or they're about the latest underground muzak sensation. A zine is not defined by its topic as its funkiness, its personality and its lack of pretension. It's easier to define what a zine is *not*. They aren't slick, they aren't professional (although some of the best publishers *do* pay a lot of attention to detail and presentation) and they don't make a profit".

GPZ falls into the genre of jobzines. The most famous of these deal with life in vulnerable portions of the service economy, zines such as *Dishwasher*, *Temp Slave*, and *Temp Slave's* predecessor, *Processed World*. The first two served as major inspirations for Helms's project:

I used them as models for the basic concept of the zine. It's a jobzine I call it. And the idea is to just assume ownership of the job and the occupation, so the debate about ethics in human research, that's for us to dictate terms. Like the scientists are on the other side of the fence, just like management's on the other side of the fence. The consumer organizations, the consumers of the data, they're all on the other side of the fence. It's our bodies and our health and safety that's at stake so we're going to dictate terms. And on that assumption, I went ahead and started the political dialog as guinea pig zero [which is both the name of the zine and his *nom d' plume*] and also I wanted to capture the culture which means get the feeling across of what it's like to be a guinea pig and go in and chew the fat with other guinea pigs around the rec room table and around the dinner table".

It is clear that the purpose of the zine is twofold: first, to discuss and promote the obscure culture of guinea pigging and, second, to bring a historical and political consciousness to the job.

The content of the zine ranges from report cards on the quality of treatment at different research units, to fiction about either guinea pig humans or rodents (usually as allegory for the humans), to poetry, to first-person accounts of experiences in research units, to editorials, to historical timelines of the history of guinea pigging, to reports on historical struggles over the job (including guinea pig strikes and anti-doctor riots) to more mundane biographies of guinea pigs of the past. The zine also has distinct a aesthetic, one close to the grotesque. To start with the stories, both historical and first-person, are filled with mishaps, abuses, and dangerous situations. Descriptions of nausea, blood, and torment, if not actual torture, are common. Some of this is part of a politics which sees the only protection for guinea pigs as coming from investigative journalism. As Helms notes, "The venue for protecting the rights of guinea pigs is going to be in scandals, exposing abuses, and, uhm, I've been constantly saying since couple of years now, investigative journalists have done more to protect human subjects by far, universes beyond what the government's done, and that's the way it's always been. It's a scandal-driven history" (1999).

On the other hand, the aesthetic of the grotesque extends beyond this journalistic function. For instance, the zine has pictures of dissected guinea pigs (the rodents) as mere decoration on the cover and in articles. There are also recipes for guinea pigs and images of eating guinea pigs, from South America where they are a foodstuff. On top of this are the stories themselves, which often include a macho element in the description of horrors. For instance, Teresa Dulse (a zine editor in her own right) submitted a story about a study she participated in, writing, "The guys claimed more nausea, more often than us. They retreated to their beds faster, too . . . I felt gross a few times. Not enough to tell the staff or to pull a trash can over to my bedside. No. 7 yacked loudly from the other room one morning . . . No. 2 was feeling so bad that he hid the pills under his tongue during the dosing procedure. The coordinator even checked his mouth and he still got away with it, spitting the pills down the toilet later on" (*GPZ #5*, 25). This is not presented as a critique. She ends the article by recommending the unit to guinea pigs in her area. There is something daredevil in the telling of these stories, of pushing the body beyond endurance.[6]

But it is the counter-histories in *GPZ* that I want to point to as a resource for educators. These histories, offering a guinea pig standpoint, challenge the ritually encoded hero tales relayed in science textbooks: Watson and Crick, Einstein and Curie, Dalton and Darwin. Science education of all flavors has tended to enshrine one subject position in the web of science as the only "real" one: that of the scientist. By putting guinea pigs back into the story, the entire structure of science education's narrative changes. Against tales of frontier heroism are tales of poor women and men whose bodies supplied the knowledge we study in science classes. It is this juxtaposition I turn to in the continuation of this thread.

World-Traveling as Another Standpoint

Haraway sheds some light on the process of taking Others' positions, which is unproblematic in Smith and Harding but is key to the standpoint methodology. What Haraway makes clear is that neither taking such positions or even taking our own can be seen as natural. "To see from below is neither easily learned nor unproblematic, even if 'we' 'naturally' inhabit the great underground terrain of subjugated knowledges".[7] But she stops there, never specifying the sort of work and technologies needed for such an inhabiting.

One alternative way of getting at this perception from elsewhere is through metaphors of travel. Philosopher Maria Lugones, for example, provides an alternative analysis to Haraway's in her discussion of "world-traveling," a form of epistemological flexibility which is "necessary for the outsider but it can also be willfully exercised by those who are at ease in the mainstream".[8] A "world" for Lugones includes such formations as communities (imagined [Anderson 1992] and otherwise), sub-cultures, occupations, races, etc. Moving between worlds is about the partial performances of self where that partiality is contested between ourselves as travelers and those whose gaze constructs us. "The shift from being one person to being a different person is what I call 'travel.' This shift may not be willful or even conscious, and one may be completely unaware of being different than one is in a different 'world' . . . One does not pose as someone else, one does not pretend to be . . . someone of a different personality or character or someone who uses space or language differently than the other person. Rather one is someone who has that personality or character or uses space and language in that way".

What this traveling accomplishes for Lugones is the interruption of arrogance as it enters into human relations and leads to the exploitation and even erasure of the Other. Beyond that, she seeks a state of playfulness or "being at ease" in particular worlds. This being at ease comes from fluency in the dialect of a particular world, agreement with the norms of the world, attachment to those in such a sphere, or through shared history. But such travel ultimately, if it strives at playfulness, interrupts arrogance. "The reason I think that travelling to someone's 'world' is a way of identifying with them is because by travelling to their 'world' we can understand *what it is to be them and what it is to be ourselves in their eyes.* Only when we have traveled to each other's 'worlds' are we fully subjects to each other" (401).

Certainly Lugones's struggle against arrogance has a bearing on science and its relationship to the world. Science, when it engages in fantasies of penetration, objectivity as a view from nowhere, and voyeurism, is the model of an arrogant world. Of course, this is not the only form of scientific practice, as Evelyn Fox Keller has documented in her biography of biologist Barbara McClintock. Science is also a practice of love, understanding, and relationship.

Such scientific world-traveling is very much the business of Haraway in her recent book.[9] Here, Haraway traces a fictive kinship system between herself, Onco-mouse (the first genetically engineered, patented organism), feminist science fiction, the homeless, scientist-activists, school teachers, and vampires, in all their historical, moral, and political ambiguity. She also extends the argument she began in "Situated Knowledges", introducing the notion of diffraction as a technology to disrupt the holisms of reflexivity. So much of Haraway's (non) standpoint theory is about accountability, i.e., that knowledge is always part of a system to which we are responsible. For feminists and other scholars engaged in social struggle, reflexivity is one way that that accountability is realized. Through confessional or sociological accountings of productions of our research, our selves, and our findings, we maintain at least the hope of vulnerability and partiality, i.e., situated knowledge.

Haraway's concern is that such reflexivity re-performs a holistic self, i.e., one that appears to lose its artifactual structure. Such reflection raises questions of what is real (How accurate did they get it?), and such questions are ultimately a distraction for Haraway's project: "Reflexivity has been much recommended as a critical practice, but my suspicion is that reflexivity, like reflection, only displaces the same elsewhere, setting up the worries about copy and original and the search for the authentic and really real" (1997, 16). Haraway's alternative, diffraction, seems to combine several practices. First, Haraway is less concerned with on reflecting the self as though it were a coherent whole and seems to attend more on the pragmatics of how we might employ or subvert science, technology, and other knowledge systems to craft more just futures. Second, the use of the word *diffraction* seems to imply the generation of multiple, displaced images, the overlaying of differently positioned accounts of possible futures as well as possible presents. "What we need is to make a difference in material-semiotic apparatuses, to diffract the rays of technoscience so that we get more promising interference patterns on the recording films of our lives and bodies. Diffraction is an optical metaphor for the effort to make a difference in the world" (16). The purpose here is to have the multiplicity of perspectives that Harding sought in strong objectivity, but not to collapse the sum into a grand and satisfying strong objectivity—after all, diffraction patterns are produced by difference, not unity. At the same time the result is a production: it loses all its photograph-as-captured-object realism.[10] Finally, the activist, progressive, socially engaged/enraged intent that inspired standpoint theory to start with is retained, in fact refocused on the goal of social change.

Diffractive Tales of Digestion of/in Science Education

Haraway has argued that representation, and all forms of storytelling, always have an instrumental, pragmatic, future-oriented dimension. "Stories are *means* to ways of living". Guinea pig stories are a means to power in the doctor-researcher/subject relationship. Science curricula can be read as stories about students' relations with nature, their bodies, and the institutions that represent nature and their bodies (medicine, the Environmental Protection Agency, Monsanto or alternatively People for the Ethical Treatment of Animals, EarthFirst, etc.) In *Modest_Witness*, for example, Haraway notes that high school science texts teach nature as a justification for extant human relations (1997, 103).

Having suggested (though not defined) what standpoint theorists argue, I wish to provide a brief example of the value of a guinea pig standpoint in science education. To do so, I want to look at the topic of digestion from several standpoints, to create a diffraction pattern of digestive stories, and present the role such stories (might) play in schools. I focus on digestion for several reasons. First, I chose it because it is seemingly such an unlikely topic for the cultural study of science, far from the charged struggles over environmental education, reproduction, and dissection. Second, I have picked it because an enormous amount is known about the history of the biology of digestion and because it is a gripping, grotesque story, the kind likely to interest junior high and high school students. Finally, I chose it because my account complements Angela Barton's analysis of digestion and the politics of poverty. The three accounts that I wish to look at are, first, the version conveyed in the National Science Education Standards, second, the version conveyed in biological and medical history texts, and, finally, digestion as seen through the recovered guinea pig histories of Robert Helms.

The *Standards'* Story

As has been discussed, the *Standards* tend to be blind to standpoints other than those that lionize scientific and medical authority. Given this, it is not surprising that despite rhetoric imploring increased concerns with history and social perspectives, digestion is placed in neither context. This does not mean that the image of digestion is without context or even uninteresting. In fact the image is quite radical from the perspective of one who was taught digestion merely as the biochemical and biophysical extraction of chemicals for the body's needs.

Digestion, as a topic, is relatively unexplored in the *Standards*, in keeping with their goal that there will be "less emphasis on knowing scientific facts and information" and more on "understanding scientific concepts and developing abilities of inquiry" (NRC, 113). It does emerge briefly in two threads of the content standards. First, it appears in the life science thread. In elementary school students are supposed to learn that air, water, and food are needs for animals. In the fifth through eighth grade portion of the *Standards*, the body is broken down into groups of tissues that serve as larger functional units, and here digestion is just briefly mentioned as the first in a list of such organ systems. The others mentioned are "respiration, reproduction, circulation, excretion, movement, control, and coordination, and for protection from disease" (p. 156). It is only at the senior high school level that digestion is dealt with specifically in this thread. "The chemical bonds of food molecules contain energy. Energy is released when the bonds of food molecules are broken and new compounds with lower energy bonds are formed. Cells usually store this energy temporarily in phosphate bonds of a small high-energy compound called ATP" (186). Strangely, the details that usually mire high school discussions of digestion, the specific series of enzymes for metabolizing carbohydrates, proteins, and fats, are absent, left to the teacher to include if they want, but not detailed in the *Standards*. Here the moral is simple: digestion is about energy. Furthermore, this detail is part of a substandard called "matter, energy, and organization in living systems" and thus places this view of food as energy in the context of ecological relationships. Digestion is thus part of a web of connections, a web whose threads consist of who is eating whom. Ecology in the form of food chains and the interspecific circulation of energy is the context of digestion. Food is not for the individual, but for the system.

The second thread concerning digestion occurs in the substandard on "science in personal and social perspectives." At various grade levels teachers are to instruct students about the relationship between nutrition and health, specifically, "eating less sugar, and eating less fat" (NRC 1996, 140). Digestion is thus linked to health, cultural norms, and, with some argument, morality.[11] This dimension of digestion, i.e., its links to health, productivity, and morality, is hardly new.

If digestion is one of the engines of ecology and a primary site of health, it is *not* an unfolding historical story here, a story about knowledge and power in science, as portrayed in the *Standards*. Nor is it a story about the consequences of getting knowledge. Nevertheless, the *Standards'* story for educators is novel in as much as it ties students to other creatures through systems of energy distribution. But digestion is also a tale of power in science, voyeuristic knowledge, and to understand this we have to listen to others' tales.

The Researcher's Story
In the late seventeenth century the degree to which digestion was mechanical versus chemical was very much a controversy in biology and medicine. In the words of physiologist Howard Haggard, "The stomach was looked upon as a mill, a fermenting vat, or a stewpan. It was [William] Beaumont who proved that digestion was a chemical process". While stationed in the Upper Peninsula of Michigan, Beaumont helped a Canadian named Alexis St. Martin, who was accidentally shot. The wound

provided a hole into St. Martin's stomach which permitting Beaumont the opportunity to experiment. This story is retold in most medical and biological histories. Beaumont is considered one of the first important American (i.e., post-Revolutionary) medical researchers. He is usually hailed as America's first great physiologist. Furthermore, it is one of the only medical stories in which the guinea pig, Alexis St. Martin, is always named in the accounting. So prominent is St. Martin's role that one book even included St. Martin's name with the doctor's in the index entry for that section. Here is a rather typical account from a medical history text: "Exactly contemporary with Drake lived William Beaumont (1785–1853), an army surgeon who treated the Canadian Alexis St. Martin for gunshot wound of the stomach. The patient recovering with a large gastric fistula, which Beaumont recognized as providing a unique opportunity for the study of digestion. At considerable personal inconvenience and expense he followed his patient in order to complete his investigations on the rate of digestion of various foods and on the composition of the gastric juice. His *Experiment and Observations on Gastric Juice and the Physiology of Digestion* (1833) recorded a fine piece of research in the face of unusual difficulties". The nature of this fine research is described by Clandening:

> [Martin St. Alexis] retained until the day of his death, sixty years later, an opening from the surface of his body directly into his stomach.
> Dr. William Beaumont . . . could see the inside of St. Martin's stomach. He could watch its movements. He could place food inside it and study the changes which occurred on the part of the food as well as the stomach wall. He showed that the stomach digests only certain kinds of foods, and that the gastric juice contains hydrochloric acid and some other substance, to Beaumont vague, but later identified by Schwann as pepsin.

Just as the *Standards*' story is not only about food but about health and interspecific relationships, these stories are not only the history of digestion as biochemical but are also tales about the heroism of Dr. Beaumont. This is most obvious in the phrase "at considerable inconvenience and expense" in Guthrie's account. Other authors emphasize the generosity of Beaumont. Kass and Kass, for instance, writes that "Alexis St. Martin . . . was befriended by Beaumont and taken into the latter's home when St. Martin was about to be returned to Canada as an indigent". Haggart emphasizes the heroism of the doctor, in his longer (two page) version but simultaneously reveals unintentionally the problems in their relationship. Haggart notes that St. Martin fled the doctor. "For two months the studies went on and then Alexis deserted Beaumont. He returned to Canada, married and had two children; there, after four years of search, Beaumont finally found him. Again, the experiments were undertaken and carried on for two years".[12]

These stories share with the *Standards'* version of digestion the deification and reification of the position of the scientist. The doctor is clever, persistent, and generous (ethical). We, like the students in the photos in the *Standards*, are invited into the position of the doctor-scientist, here through phrases that allow us to experience the world from Beaumont's point of view: his frustrations, his sense of being deserted, and his largesse. At the same time, it introduces a historical view, and begins, however hesitantly, to make clear that the knowledge of digestion emerges through historically unfolding human relations, not outside of them.

The Guinea Pig's Story

Robert Helms in *Guinea Pig Zero* #6 renarrates the St. Martin–Beaumont episode, but he organizes it around St. Martin's life: his reasons for being in the Upper Peninsula (he was hired by the American Fur Company as a *voyageur* or sales representative) and his reasons for leaving Beaumont when he did (they had moved around the country for nearly a year together and were close to Montreal at the time he decided to leave). From this guinea pig stance, new questions get raised about the line that Beaumont crossed in moving from doctor to scientist. Helms points out that after it was clear that St. Martin was going to survive the gunshot he could have performed "another operation to close up the hole and separate the stomach from the body wall" (8). His failure to do so in fact was controversial even in his day, and his ethics were publicly debated in 1834, when he sought government funds for his research, and then later (1840) when he, among other surgeons, was charged in the murder case of an assaulted St. Louis politician who died on the operating table. Beaumont, for his part, decided to bore a hole in the poltician's skull to relieve the pressure. In his trial the prosecutor argued that "it was upon the same principle of curiosity which kept the hole open in the man's stomach that he bored a hole in Davis' [the politician] head to see what was going on in there." (9). Helms is thus not anachronistically applying a moral code when he writes, "It is difficult to suppose that the physician did not actively steer the course of events in such a way that the living body of the young fur trader would be of maximum benefit to his own career. This would be almost the same as explaining why a surgeon might perform a heart operation and then leave the patient's chest open in order to watch the heart beating for the following sixty years . . . *This* patient was denied the final and most obvious part of the treatment for his injury, and therefore Beaumont's use of Alexis St. Martin as an experimental subject was exploitative and unethical" (9).

Helms provides St. Martin with a broad, multifaceted biography. He goes on and describes the three stretches when St. Martin served as a guinea pig for Beaumont, their complicated relationship in which St. Martin felt both indebted to the doctor and resentful. He describes St. Martin's marriage and how that influenced his comings and goings; his wife, Marie Joly, and their children (six in all) accompanied St. Martin when he returned to the Midwest to serve as a guinea pig in the second and third group of experiments. And Helms describes the economics of their relationship: the contracts they signed, the agreements Beaumont struck with the army and the fur companies to support St. Martin and his family. He also describes how St. Martin came to tour the United States and Europe and was publicly exhibited, first, by Beaumont and later, after Beaumont died in 1853, by a fraud called Dr. Bunting. Alexis St. Martin died in Canada in 1880. Various doctors and institutions tried to obtain or purchase his "famous stomach" (*GPZ #6*, 15) posthumously, including William Osler, Johns Hopkins University's first medical professor, for the Army Medical Museum in Washington. "Relatives . . . kept Alexis' body at home much longer than usual during a hot spell of weather so as to let it decompose as much as possible and be of as little use as possible to science. Also they dug the grave eight feet below the surface instead of six to prevent 'resurrectionists' in the employ of doctors from robbing the corpse" (15).

While this guinea pig story shares with the doctor's a biochemical (versus ecological) view of digestion, it departs from it in a number of critical ways. Most importantly, it accomplishes both Haraway's goal of partiality and Lugones's goal of anti-arrogant world-traveling. In Helms's version of the digestion story, partiality plays multiple roles: we are made aware that being a guinea pig was a smaller part of the greater ineffable whole of St. Martin's life. We are also made aware of the limited beneficence of scientific curiosity and of the partiality of the doctor's stories—of the ways in which St. Martin's life and interests have been pushed off stage, as necessary for the god-trick to be convincing and for Beaumont to carry the full burden of the hero's role. But just as critical are the ways that the guinea pig story works in every phase of its telling to indict the arrogance of the scientific establishment, by forcing us to take the standpoint of the research subject. This critique is not subtle or left to the reader's imagination. At one point, after reprinting a letter from St. Alexis to Beaumont, Helms notes that attached to the letter was a comment by the University of Chicago library curator, Arno Luckhart, which ends, "Alexis St. Martin will continue to engage the attention of posterity because of the genius of Beaumont" (p. 12). Helms comments: "Luckhardt's comments are . . . a fine example of a person of privilege whose compulsion to abuse working people will emerge from him at every point . . . For another thing, it shows the way human research subjects have been viewed always: as the servants of their learned manipulators, who are human only in the mechanical sense, and who should be grateful to be in such distinguished company on any terms" (12). It is the ability to witness this arrogance that the guinea pig standpoint adds to our knowledge of science.

Guinea Pig Pedagogy

As science teachers narrate nature to their students, they produce standpoints from which that nature is apprehended. However, with few exceptions, teachers have presented only the imaginary standpoint from which scientists appear as heroic, (im)modest witnesses to the workings of the world. What I am advocating in spinning these multiple stories is that teachers and researchers need to weave these and other standpoints into our research, standpoints which check the arrogance of knowledge and standpoints that students are more likely to occupy in their life courses. These stories can help students understand the possibilities of agency and suffering for portions of lives lived in these locations.

But the inclusion of these stories is just a starting point. The stories alone just hint at what it means to teach through other standpoints. The point of including the story of Alexis St. Martin, for example, is not one of warning or morality, but for students to explore the power they might have in medical and scientific encounters; and such exploration requires more than just listening and telling. Thus, just as in current scientist-centric pedagogy, it is insufficient to hear stories of heroism (e.g., about Einstein or Watson) and learn facts. Students must put on the lab coats and play with animals, projectiles, and chemicals, thereby learning to treat the world as material objects subject to the scientist's gaze; so in a guinea pig pedagogy one must engage in imaginative exercises (research, role playing, autobiography) to see oneself as a guinea pig and to understand what dangers and possibilities lie in being the object of science.[13] Having students travel, in Lugones's sense of the word, to these subject positions in meaningful ways is what defines a guinea pig pedagogy.

Conclusions: Guinea Pig Pedagogy Is Diffractive Pedagogy

Finally, my argument here is *not* that we should replace the ecologist's (the *Standards'*) tale or the doctor's tale with the guinea pig's tale in science pedagogy, i.e., substitute one monocular truth for another. To do so would lose the partiality of Helms's own stories. We are better served by insisting that science educators learn to tell multiple tales from inside the web of scientific practice, learn to travel between the roles of scientist-doctor, guinea pig, citizen-participant, janitor, spouse, etc. Elsewhere, I have called this approach to science education multifocal, in its insistence on multiple, non-synthesized (in the Hegelian sense) standpoints and identifications for students, identifications with the powerful and the subaltern. But this unsettling multiplicity also points to it as a form of pedagogical diffraction.

Clearly this approach owes much to all three of the theorists indexed above: Harding, Haraway, and Lugones. From Harding I have taken the idea of focusing on knowledge that has emerged out of social struggle, i.e., the struggle of human subjects to have rights in the medico-scientific encounter. My emphasis on partiality and multiplicity as well as my legitimization of the guinea pig position itself—one that is distinct from, though crosscut with, the more traditional ones produced in race, gender, and class relations—comes from my reading of Haraway. Finally, the focus on a pedagogy which challenges arrogance through world-traveling and exploring difference comes from Lugones's metaphors of mobility

Guinea Pig Zero provides the biology classroom with one easily available, alternative perspective to the hero worship normally proffered by texts and other extant science education resources. It reminds students of the price of scientific of knowledge in terms of human life and also offers the possibility for politics within science for the times when they find themselves on the subject end of the scientific relationship. But including the guinea pig perspective is only a first step towards having the students explore the many power-charged subject positions that are science and for which we must educate.

Acknowledgments

I want to thank Robert Helms for his time and generosity, the attendees of the CASTAC 1999 workshop for their important insights into guinea pig anthropology, David Shutkin for various leads, and Angela Barton and Margery Osborne for their leadership, thought, and struggles.

To obtain *Guinea Pig Zero* contact
  GPZ
  PO Box 42531
  Philadelphia, PA 19101
  gpz@netaxs.com or http://www.geocities.com/hotsprings/villa/2529/

References

Altman, Lawrence K. 1986. *Who goes first: The story of self-experimentation in medicine.* Berkeley: University of California.

Anderson, Benedict. 1992. *Imagined communities.* 2nd ed. London: Verso.

Apple, Michael W. 1990. *Ideology and curriculum.* 2nd ed. New York: Routledge.

Bakhtin, Mikhail. 1988. *Rabelais and his world.* Translated by H. Iswolsky. Massachusetts: MIT Press.

Bakhtin, Mikhail Mikhailovich. 1981. *The dialogic imagination: Four essays.* Translated by M. Holquist. Austin: University of Texas Press.

Barthes, Roland. 1977. *Image–music–text.* Translated by S. Heath. New York: Hill and Wang.

Barton, Angela Calabrese. 1998. Reframing "science for all" through the politics of poverty. *Educational Policy* 12 (5):525–541.

Bowers, C. A. 1992. The relevance of traditional (ecologically sustainable) cultures for thinking about deep cultural approaches to environmental education in urban American classrooms: Exploring the distinction between 'copying' and 'learning from'. Paper read at the meeting of the Journal of Curriculum Theory (JCT), October, at Dayton, OH.

Carlsson, Chris, and Mark Leger, eds. 1990. *Bad attitude: The Processed World anthology.* New York: Verso.

Clendening, Logan. 1933. *Behind the doctor.* New York: Alfred Knopf.

Clifford, James. 1997. *Routes: Travel and translation in late twentieth century.* Cambridge: Harvard University Press.

Damarin, Suzanne K. 1993. Schooling and situated knowledge: Travel or tourism. *Educational Technology* (January).

Douglas, Mary. 1966. *Purity and danger: An analysis of concepts of pollution and taboo.* New York: Praeger.

Eisenhart, Margaret, Elizabeth Finkel, and Scott F. Marion. 1996. Creating the conditions for scientific literacy: A re-examination. *American Educational Research Journal* 33 (2):261-95.

Evans, Gavin. 1999. Health: Guinea pigs in the middle;. *The Guardian*, January 19, 1999, 14.

Gabriel, Jospeh M. 1998. Treadmill of history. *Guinea Pig Zero* (5):13.

GPZ. 1996. Treadmill of history. *Guinea Pig Zero*, May, 1-4.

Guinea Pig Zero (GPZ) 1995-2000. Edited by R. Helms. Philadelphia.

Guthrie, Douglas. 1946. *A history of medicine*. Philadelphia: J. B. Lippincott.

Hacking, Ian. 1983. *Representing and Intervening: Introductory topics in the philosophy of natural science*. Cambridge: Cambridge University Press.

Haggard, Howard W. 1934. *The doctor in history*. London: Oxford University.

Haraway, Donna J. 1989. *Primate visions: Gender, race and nature in the world of modern science*. New York: Routledge.

———.1991. The biopolitics of postmodern bodies: Constitutions of self in immune system discourse. In *Simians, Cyborgs, and Women: The Reinvention of Nature*, edited by D. J. Haraway. New York: Routledge.

———. 1991. A cyborg manifesto: Science, technology, and socialist-feminism in the late twentieth century. In *Simians, Cyborgs, and Women: The Reinvention of Nature*, edited by D. J. Haraway. New York: Routledge.

———. 1991. Situated knowledges: The science question in feminism and the privilege of partial perspective. In *Simians, Cyborgs, and Women: The Reinvention of Nature*, edited by D. J. Haraway. New York: Routledge.

———. 1994. A game of cat's cradle: Science studies, feminist theory, and cultural studies. *Configurations* 2 (1):59-71.

———. 1997. *Modest_Witness@Second_Millennium.FemaleMan©_Meets_OncoMouse™.* New York: Routledge.

Harding, Sandra. 1991. *Whose science? Whose knowledge? Thinking from women's lives.* Ithaca: Cornell University Press.

Helms, Robert. 1998. Alexis St. Martin: The intrepid guinea pig of the Great Lakes. *Guinea Pig Zero* (6):6–16.

———. 1999. Interview by Matthew Weinstein. Tape Recording. Philadelphia, PA, 29 March.

Hurston, Zora Neale. 1990. *Mules and men.* 1st Perennial Library ed. New York, N.Y.: Perennial Library.

Jones, James H., and Tuskegee Institute. 1993. *Bad blood: the Tuskegee syphilis experiment.* New and expanded ed. New York: Free Press.

Kass, Amalie M., and Edward H. Kass. 1988. *Perfecting the world. The life and times of Dr. Thomas Hodgkin, 1798–1866.* Boston: Harcourt Brace Jovanovich.

Keller, Evelyn Fox. 1983. *A feeling for the organism: The life and work of Barbara McClintock.* New York: W. H. Freeman.

Kelly, Jeff. 1997. *Best of Temp Slave!* 1st ed. Madison, WI: Garrett County Press.

Kondo, Dorinne K. 1990. *Crafting selves: Power, gender and discourses of identity in a Japanese workplace.* Chicago: University of Chicago Press.

Latour, Bruno. 1987. *Science in action.* Cambridge: Harvard University Press.

Longino, Helen E. 1993. Feminist standpoint theory and the problems of knowledge. *Signs* 19 (1):201–12.

Lugones, María. 1990. Playfulness, "world"-traveling, and loving perception. In *Making face, making soul, hacienda caras: Creative and critical perspectives by women of color,* edited by G. Anzaldúa. San Francisco: Aunt Lute Foundation Books.

Monmaney, Terence. 1999. U.S. suspends clinical research at Duke. *Los Angeles Times,* May 12, A-18.

NRC. 1996. *National science education standards.* Washington, DC: National Academy Press.

Olson, Gary A., and Elizabeth Hirsh, eds. 1995. *Women writing culture.* New York: Suny.

Pauly, Philip J. 1996. The development of high school biology: New York City, 1900–1925. In *The Scientific Enterprise in America: Readings from Isis,* edited by R. L. Numbers and C. E. Rosenberg. Chicago: University of Chicago.

Pore, Jerod. 1999. *General information about zines and alt.zines* [www]. Factsheet 5 1995 [cited July 1999]. Available from// www.factsheet5.com/faqs.html.

Rodriguez, Alberto J. 1997. The dangerous discourse of invisibility: A critique of the National Resource Council's National Science Education Standards. *Journal of Research in Science Teaching* 34 (1):19–37.

Rowe, Chip. 1999. *The book of zines* [www]. Chip Rowe 1997 [cited July 1999]. Available from http://www.zinebook.com/whatcha.html.

Shaw, Jane, and Michael Sanera. 1996. *Facts, not fear: A parent's guide to teaching children about the environment.* Washington, DC: Renery Publishing.

Silin, Jonathon G. 1995. *Sex, death, and the education of children: Our passion for ignorance in the age of AIDS.* New York: Teachers College Press.

Smith, Dorothy E. 1987. *The everyday world as problematic: A feminist sociology.* Boston: Northeastern University.

Stallybrass, P., and A. White. 1986. *The politics and poetics of transgression.* Ithaca: Cornell University Press.

Teegardin, Carrie. 1997. Veterans used as guinea pigs, VA doctors say; 2 researchers may have coerced patients. *St. Louis Post-Dispatch,* June 8, A-4.

Trafford, Abigail. 1996. Bitter medicine. *Washington Post,* July 2, 1996, Z-6.

Traweek, Sharon. 1996. Unity, dyads, triads, quads, and complexity: Cultural choreographies of science. In *Science Wars,* edited by A. Ross. Durham, NC: Duke University Press.

Tsing, Anna Lowenhaupt. 1993. *In the realm of the diamond queen: Marginality in an out-of-the-way place.* Princeton, N.J.: Princeton University Press.

Weinstein, Matthew. 1998. Playing the paramecium: Science education from the stance of the cultural studies of science. *Educational Policy* 12 (5):484–506.

———. 1998. *Robot World: Education, popular culture, and science.* New York: Peter Lang.

———. 1999. A public culture for guinea pigs. Paper read at the meeting of the Committee for the Anthropology of Science, Technology and Computing, June, at New York, NY.

Notes

1. It should be noted that medical personnel, including medical students, nurses, and doctors themselves, have also always been subjects; both for reasons of convenience and of ethics.

2. The embrace of this term by human subjects has only followed the modernization of the process. Subjects of the Tuskegee experiment, for instance, vehemently distanced themselves from the term.

3. A short list of such groups would include the American Association for the Advancement of Science, the National Science Teacher's Association, the National Research Council, the American Chemical Society, and the American Association of Physics Teachers.

4. The Belmont Report was carried out by the National Commission for the Protection of Human Subjects of Biomedical and Behavioral Research as charged by the 1974 National Research Act.

5. Of course there are different portions of the guinea pig labor market: there are people with diseases for whom drug trials may be the only hope of life or relief. There remain the doctors and other medical professionals who see their own self-experimentation as a form of ethical practice, and then there are the healthy guinea pigs who sign up for studies of side effects or for general psychological experimentation. While rules concerning consent apply across all these groups, it is the last group that is under discussion here.

6. Drug use is an issue here also. While Helms notes that the reputation for drug use among the guinea pigs is exaggerated, since the subjects must have high-functioning livers to get into studies, the guinea pig community has a large volume of lore about masks (teas and other concoctions) for hiding drug residues in the blood. This should not be surprising, the willingness to expose oneself to dangerous substances is a major criterion of the job. In this way the economy of science depends on licit and illicit drug use.

7. Haraway does not separate theory, perspective, standpoint, and narrative from each other as Smith and Harding are at pains to do. For Haraway storytelling, theorizing, and envisioning are all complicated, technically assisted acts of world-framing. For Harding and Smith the distinctions are critical to prevent naturalizing the accounts of women in the world and thus having standpoints represent the sort of organic wholes which are never present in Haraway's account.

8. Suzanne Damarin has tied together many of these same elements (Lugones, Haraway, and travel metaphors) along with psychological theories of situated learning in her analysis of pedagogy and technology. She uses the travel (or geographical displacement) metaphor to get at the quality of learning, noting that modes of being in the Other's location include, in addition to travel, tourism. To

her list I would add colonialism and coercive (e.g., military) occupation. These alternative displacement metaphors play out in curriculum and pedagogy when we erase, exploit, etc., both absent and present lives.

9   Haraway's travels in this text are in part anthropological. She claims to have applied for "a visa for an extended stay in the permeable territories of anthropology" (49). Anthropology has been a sustaining form of "world-traveling" for white (and, more recently, Others as well) academics, though one prone to arrogance in its committment to colonialism, social control, and stereotypification. At the same time, there have always been counter streams of playful world-traveling in anthropology, e.g., the work of Zora Neale Hurston, Anna Tsing, and Dorrine Kondo. Even the realization of James Clifford of traveling cultures—that the Others we study may be more cosmopolitan than ourselves—unravels the arrogance of those world-traveling colonists who studied but rarely came to know the Others around them.

10 These theoretical struggles are in some ways unnecessary. The lesson of situated knowledge was, after all, that all accounts, i.e., perspectives, are constructed (artifactual, not artificial) and constructed purposefully. This is as true of reflection as of diffraction. For example, a group of teen girls I was studying at a science museum insisted on being photographed in carnival mirrors so that they could see themselves as thin and long haired. This play with glamour was clearly not the same being reflected elsewhere, but it was a reflective act, literally.

11 The moral dimension of food and eating is hardly subtle in middle-class American culture. Gluttony is held to be a sin, and whole food groups are cast as sinful and indulgent.

12 Note the selective use of first and last names to establish a status hierarchy between Beaumont and St. Martin.

13 Students are likely to have already been guinea pigs in various experiments in educational practice, such as cooperative learning, standardization, and inquiry. Classes might use education, therefore, as a starting point for a collective reflection on the meanings and consequences of experimental and scientific practice.

Chapter 10

Promoting Inclusive Science Education through Professional Development: Challenges Faced in Transforming Content and Pedagogy

Julie A. Bianchini
Lynette M. Cavazos

Current efforts to reform science education promote scientific literacy for all students, particularly for female and ethnic minority students usually positioned by educational practices on the margins of science (AAAS 1990, 1993; Ginorio 1995; Hurd 1998; NRC 1996; NSF 1996; NSTA 1992). To help meet the scientific literacy needs of all students, science education reformers recommend science teachers participate with colleagues and professional developers in regular, repeating activities that promote the critical exchange of ideas and information and that nurture the transformation of subject matter and pedagogical practices (Brickhouse 1994; Kreinberg and Lewis 1996; Nieto 1999; Rosser 1991, 1997). As professional developers, the authors of this chapter attempt to assist practitioners in the complex and difficult process of exploring, constructing, and practicing how to provide all students access to challenging, relevant science content and equal opportunities to understand scientific knowledge, skills, practices, and applications. Lynne Cavazos founded and facilitates the Women Educators of Science and Technology, or WEST, group, a collective of women secondary science teachers interested in transforming their practice. Julie Bianchini is a member of a team implementing the Promoting Women and Scientific Literacy project, an initiative undertaken by science, women's studies, and science education professors to make their undergraduate science courses more inclusive. We believe real change in science education will not occur unless and until extended professional development opportunities are made available to all science instructors—elementary teachers through university scientists—to discuss, develop, and enact science content and instruction that is inclusive, equitable, and excellent for all students, not just a select few.

In this chapter, we examine our own efforts as professional developers to support practicing science teachers in crafting and integrating inclusive *content* and *pedagogy* into their educational practices. We begin by presenting a case study of Denise, a science teacher attempting to implement female friendly, culturally inclusive curriculum and instruction in her junior high school science classroom[1]. We then describe the conceptual foundations and organizational structures of our two professional development projects. Each project strives to encourage secondary science teachers or university scientists to reflect upon their values and beliefs related to the teaching and learning of science and to move beyond awareness of inequities to the implementation of content and strategies inclusive of all students. Next, we return to the accomplishments fashioned and challenges faced by science teachers and scientists involved in our projects: Drawing from individual and group interviews, we present three additional case studies of participants who have worked to create and implement both inclusive science content and instructional practices. From these cases, we argue that curricular and instructional transformation is a complex and challenging process, one requiring teachers to construct individual visions of inclusive science education responsive to their personal interests, students' needs, and institutional demands. We also recommend professional developers support teachers during this process by providing multiple strategies for locating and integrating inclusive science content; opportunities to work collaboratively with colleagues; and time necessary to establish professional communities if inclusive science education is to become a reality in schools.

The Case of Denise

Denise[1] has been a member of the Women Educators of Science and Technology, or WEST, group since its inception in 1996. A European American woman, she is in her fifth year of teaching science at an urban junior high school in Southern California. Her undergraduate degree is in environmental studies with an emphasis in ecological balance and wildlife management. At the core of Denise's science teaching philosophy is the importance of all students making connections with both the content and process of doing science. Based on her reading of brain research (Sylwester 1995) and multiple intelligences (Gardner 1993), she views learning as a process of building bridges between new and prior understandings: "The way people learn is to attach information to something they know and understand." She strives to achieve this goal by integrating subject matter within science disciplines and across diverse content areas and by implementing teaching strategies that provide all students with multiple opportunities to understand science concepts and make personal connections to their own lives. She explains that "anytime you integrate the curriculum, the content becomes more relevant for students. Students have different strengths and abilities and an integrated approach to teaching and learning provides multiple ways for students to connect and demonstrate their understandings."

For Denise, an inclusive science unit contains integrated content and multiple active learning strategies. It begins with a central theme, or concept, and includes lessons, or learning experiences, that emanate from the center to form a complex web. The interweaving of science content and instructional strategies provides students with a variety of ways to make connections to the material and values the experiences, cultural backgrounds, and personal insights students bring to the learning environment. In the following description of her unit on "Light and the Electromagnetic Spectrum," Denise explains how she interweaves content and pedagogy to provide all students access to the science of light, color, and human sight, as well as equal opportunities to be actively involved in a community of learners:

> The unit begins with students making simple oscilloscopes that change the sound waves of their own voices into light waves using reflective sunlight on the school walls. The topic of visible light leads into lessons on the human eye as a sensory organ. Students write an essay on how their lives would change if they lost their eyesight, complete several vision tests on their own eye sight, and learn the structure of the eye before dissecting a cow's eye. During the dissection, the lens of the eye raises many questions from students providing a natural transition into learning about optical instruments. Microscopes are used to observe a bioluminescent single-celled algae, *Pyrocystis*, and each student is given a vial of algae to care for at home. At night, students are able to observe the bluish-green glow of the algae and are encouraged to explain the phenomenon of bioluminescence to their families. To learn more about color, the art teacher is invited to work with students to make color wheels and discuss the specifics of color, hue, pigment, intensity, shade, tone, and tint. The unit culminates with our tie-dye shirt lab, a trend from the 60s that is now the in-style for middle school students. The students learn about fiber reactive dyes and their reaction with cellulose molecules

in cotton shirts. Learning about pigment mixing, and primary and secondary colors of light helps make learning about light, color, and visible pigments exciting for all students and everyone comes away with a distinctive, personally designed tie-dye T-shirt.

Using an integrated approach to science teaching and learning is challenging for Denise; her challenges provide teachers and professional developers insight into the kinds of information, support, and guidance required to implement science education inclusive of all students. For example, Denise and her students find it difficult to teach and learn through integrated instruction because the process is nonlinear, rather than incremental or hierarchical as in traditional units. Denis must carefully consider how to make important connections among concepts visible for herself and for her students. She explains, "I have to think about how to move from one topic to another, for example, how to help students make the connection between bioluminescence in algae and the optical properties of cows' eyes." Denise also struggles to share these kinds of integrated units with other science teachers at her school. The interweaving of content from various disciplines is specific to her science background and ways of thinking about the interconnectedness of science: "This is how I think so it is not as difficult for me, but it could be for others. I need to see the whole picture first and then move to the details." Most other science teachers, for example, do not have an in-depth understanding of biology, chemistry, art, and history necessary to implement her unit on light and the electromagnetic spectrum. Assessment of student understanding of integrated units continues to be problematic for Denise as well because learning occurs at multiple levels in and out of the classroom setting. How does one evaluate all of the learning that occurs during a tie-dye t-shirt activity?, Denise asks. How does one craft a comprehensive assessment of unit ideas and processes?

The current movement toward "standards-based education" in the State of California raises additional challenges for Denise. Denise views the recently adopted California *Science Content Standards Grades K-12* (California State Board of Education 1998) as content heavy and as emphasizing student memorization of scientific facts and terminology over conceptual understanding. She explains that "in her department, teachers are primarily lecturing in order to cover the extensive content. They are not personalizing the material for the students because the focus now is on preparing students to perform well on standardized tests." Given the number and specificity of the standards, Denise struggles to create learning experiences that will help prepare her students for state standardized tests while continuing to utilize an integrated approach to teaching and learning; she finds it difficult to balance the requirements of state standards with her desire to provide all students opportunities to acquire scientific understanding and find personal, relevant connections to their own lives. Ultimately, Denise needs help from colleagues and professional developers to answer her most important question: How can one achieve the goals of standards-based science education while implementing an integrated approach to the teaching and learning of science?

Conceptual Framework

With Denise's instructional efforts and needs in mind, we return to the purpose of this paper: to examine challenges faced by science teachers and scientists intent on implementing gender equitable and culturally inclusive content and pedagogy in their science courses, and to contribute to the growing discussion of ways professional developers can assist practicing educators in building a more just and equitable science education. Our professional development projects and study of participants' experiences are informed by two areas of science education and teacher education scholarship: models of inclusive science curriculum and instruction, and models of professional development that move teachers beyond awareness of issues to action. Below, we briefly examine each of these areas in turn.

Inclusive Curricular and Instructional Models

As one dimension of our conceptual frame, we examine curricular and instructional models designed to make science more attractive and inviting to all students, particularly women and people of color (Atwater 1996; Banks 1995, 1999; Barton 1998; Gaskell and Hildebrand 1996; Mayberry 1998; Mayberry and Rose 1999; McCormick 1994; Middlecamp and Baldwin 1995; Middlecamp and Subramaniam 1999; Nieto 1996; Price and Ball 1998; Rodriguez 1998; Rosser 1991, 1995, 1997). Such models are diverse in conceptual frame, intent, and strategies recommended. Jeremy Price and Deborah Ball, for example, advocate the enactment of liberatory pedagogy in mathematics classrooms. They recognize the existence of multiple liberatory pedagogies rooted in radical, critical, feminist, and/or postmodern theories of education and note that "perspectives on liberatory education vary in the emphasis they place on issues of empowerment, social change, access to knowledge, racial or patriarchal oppression, and equity in schooling" (1998, 256). Price and Ball's particular version of liberatory pedagogy includes raising questions about mathematics as a discipline; interrogating the role of mathematics as presented in schools; and attending to relationships across mathematics, learners, and teachers within the context of gender, class, and race relations. They struggle to validate the knowledge and experiences of students often marginalized in the larger society, as well as to teach these same students standard academic knowledge. In other words, they attempt to enact liberatory pedagogy by disrupting power relations within the mathematics classroom while providing opportunities for students to access and master the discourse of those in power.

Catherine Middlecamp and Banu Subramaniam define feminist pedagogy as sharing "its roots with alternative pedagogies" like liberatory pedagogy but "distinct in its focus on women and their experiences both in and out of the classroom" (1999, 520).[2] A scientist and women's studies scholar respectively, they view feminist pedagogy not as prescribing a series of formulas for teachers to implement, but as offering teachers ideas to inform their own instruction given particular teaching styles, student needs, and institutional constraints. In other words, they understand feminist pedagogies to "enhance our ability to use our individual ways of teaching to promote student interest and learning" (1990, 521). Although they avoid specific directives, Middlecamp and Subramaniam identify several salient themes held across diverse feminist pedagogies: These themes include recognizing women's lived experiences and interrupting unequal gender relations; fostering the existence of multiple authorities in the classroom; presenting people and knowledge as positioned within and across different contexts; empowering students to draw from their own resources, utilize their own strengths, and serve as their own mentors; helping students find and fashion their voices, for example, by presenting material that is relevant and connected to their lives; and finally, challenging claims that scientific knowledge is free of values or interests.

Middlecamp and Subramaniam (1999) also make clear that inclusive science education speaks both to *what* is taught and *how* it is taught—to both science content and pedagogy. Feminist, scientist, and science educator Maralee Mayberry (1998) stresses this same point: Both curriculum and instruction must be transformed if science education is to address the interests and experiences of all students. Innovative techniques are necessary to restructure the environment within which students learn science, she argues, but are not sufficient to empower students to change the norms and practices of science or the larger social order within which science is embedded. To create a truly transformative science education,

Mayberry explains, attention to science content that articulates students' experiences and concerns is needed as well. Mayberry offers her own interdisciplinary undergraduate course, Earth Systems: A Feminist Approach, as a discipline-specific model of gender equitable, culturally inclusive science education. In a unit on petroleum resources and reserves, she implements a simulation where students act as oil companies purchasing land and drilling for oil. Within the context of this innovative instructional strategy, students are provided access to transformative science content: They both learn about the geological concepts of sedimentary strata, faults, folds, and natural resources and gain insights into how the discipline of geology intersects with the natural environment, economic interests, and political forces.

In addition, Sue Rosser (1991, 1995, 1997), James Banks (1999), and Sonia Nieto (1996) remind teachers and professional developers that the creation and enactment of inclusive content and pedagogy takes time; each provides a model of curricular and pedagogical transformation that progresses in discrete stages or levels. Rosser's (1991, 1995, 1997) female friendly model of curriculum and instruction, for example, targets science and mathematics education at the undergraduate level. In Rosser's model, practitioners move from phase 1, where the absence of both women in the scientific enterprise and women's issues in the science curriculum is not recognized, to phase 3, where barriers that prevent women from pursuing science and strategies found successful in eliminating them are identified, to phase 6, where science is redefined and science education, reconstructed to include all. In a second model, Banks (1999) identifies four approaches to address the learning and cultural styles of diverse student groups in any discipline across K-16 (kindergarten through college) education. The first and second approaches, the rudimentary and additive approaches, Banks explains, are multicultural in intent but fail to challenge the norms and values of the dominant culture reflected in the curriculum. In contrast, the fourth approach, decision-making and social action, empowers students to consider concepts, events, and people from diverse perspectives; understand knowledge as socially constructed; develop skills to analyze, formulate, and justify conclusions; and pursue projects which require personal, social, and civic actions related to the ideas and issues studied. Nieto (1996) presents a third model for inclusive and comprehensive multicultural education in K-12 settings. Like Banks, she identifies four levels of support for pluralism within the school setting: tolerance; acceptance; respect; and affirmation, solidarity, and critique. As educators move through these levels, Nieto explains, they are challenged to confront the inequalities that exist within their predominantly monocultural schools; incorporate inclusive curricular materials and instructional strategies that affirm, support, and celebrate differences in students and their families; and work collectively with students toward the goals of social justice.

Models of Professional Development

As stated above, central to Rosser (1991, 1995, 1997), Banks (1999), and Nieto's (1996) models of inclusive curriculum and instruction is the notion of phases or levels, the recognition that practitioners must engage in a lengthy, complex, and iterative process of curriculum development and instructional improvement if science content and pedagogy are to be transformed. To support science teachers in these efforts, professional developers must create opportunities for teachers to grow personally and professionally, deepen their understanding of the needs and interests of their own students, and craft inclusive curricular and instructional strategies responsive to local institutional demands. Thus, the second piece of our conceptual frame is fashioned from models of professional development that support the goal of inclusion (Kreinberg and Lewis 1996; Loucks-Horsley, Hewson, Love, and Stiles 1998; Muller and Pavone 1998; Sanders, Campbell, and Steinbrueck 1997). These models were used to inform the design and implementation of our own inservice projects for practicing science teachers or university scientists.

Susan Loucks-Horsley, Peter Hewson, Nancy Love, and Katherine Stiles (1998), for example, recommend three organizing principles be used to craft professional development programs focused on equity and access for all students. These first principles of professional development underscore the importance of valuing each and every teacher's knowledge and experiences: providing access for all teachers to quality professional development initiatives; designing programs to accommodate the diverse characteristics of teachers; and ensuring that what teachers learn provides them with the skills, resources, and sensitivities necessary to help a diverse student body gain literacy in science. Equally important, Loucks-Horsley et al. continue, professional development opportunities must take the form of extended and sustained efforts to build professional communities, rather than isolated and occasional teacher workshops. A strong professional culture is essential to support exploration and promote innovation—to help science educators move beyond mere recognition of issues of gender and ethnicity to initiate changes in norms of practice and inclusive pedagogy.

Nancy Kreinberg and Sue Lewis (1996) draw from their own experiences with gender equitable and/or culturally inclusive professional development initiatives to fashion a second series of recommendations. From their work on the EQUALS program and McClintock Collective respectively, Kreinberg and Lewis begin by asking professional developers to recognize that there is no one right way to educate all students. Teachers must be viewed as advocates of equity, allowed the freedom to develop methods responsive to their own students' needs, and trusted to do what is right and equitable, they argue. Like Loucks-Horsley et al., Kreinberg and Lewis recommend professional developers provide a forum for teachers to work together to critically examine education reforms and to see in what ways they can effect change: "Teachers need to be able to see how what they do makes a part of the whole and that they have a potential, with others, to transform what no longer is useful into what is needed" (1996, 189). Finally, the pair urge professional developers to press for changes in the school system as a whole so that their work with collectives of teachers around issues of equity can thrive. Thus, they challenge professional developers to design opportunities for teachers both to explore equitable practices and to identify strategies for creating alternative ways of schooling.

Methodology

Context of Study: Professional Development Projects

Data for this study were drawn from two professional development projects designed to promote extended examination of issues of gender and ethnicity in science; these two projects share several principles with professional development models discussed in our conceptual framework above. The Women Educators of Science and Technology, or WEST, group consists of women secondary science teachers who graduated from the Teacher Education Program at the University of California, Santa Barbara and who meet regularly to explore ways of making their curriculum and pedagogy more inclusive. The WEST group was founded by Lynne Cavazos, a faculty member of this teacher education program, in 1996. The Promoting Women and Scientific Literacy project is an initiative undertaken by scientists, women's studies faculty, and science educators at a public, urban university to transform undergraduate science education. The project is coordinated by the Association of American Colleges and Universities (AAC&U) and funded by the National Science Foundation (NSF). Julie Bianchini led one aspect of this multiyear, multipronged effort: a professional development seminar series designed to assist university scientists in making their introductory science courses more inclusive of women and ethnic minorities. We discuss the goals, participants, and approaches of these two professional development programs in greater detail below; each description is written in the first person.

Lynne Cavazos and Women Educators of Science and Technology. In their work on narrative inquiry, Jean Clandinin and Michael Connelly (1990) emphasize the importance of teachers finding their voices and exploring their beliefs about teaching and learning within a community setting. Using this as a guiding principle, the WEST group was created to provide a space for women teachers to collectively examine science and science education from their perspectives as women, scientists, and teachers. The notion of perspectives is central to the group. WEST members see themselves and their science students as coming to "know science" through personal experiences. They recognize that teachers and students acquire a particular image of science from those life experiences they find memorable and noteworthy, the positive as well as the negative.

To understand the origin and organization of the WEST group, I begin with my own professional journey from science teacher to science teacher educator. During the first 20 years of my secondary science teaching career, as a European American woman, I never questioned my traditional understandings of scientific knowledge or the underlying biases of our culture regarding objectivity, truth, and the scientific enterprise. Why should I? I was trained to be a science teacher by completing course and lab requirements in the disciplines of science and by modeling my teaching after male science professors and their male teaching assistants. As a classroom teacher, my questions focused on issues of depth versus breadth of content, access to materials and equipment for student use, and implementation of activities and projects that would connect students to science in their own lives. It was not until my doctoral program, as a member of the WISE group (Women's Issues in Science Education), that I was introduced to the work of feminist scholars such as Ruth Hubbard, Evelyn Fox Keller, Carol Gilligan, and Mary Belenky. As a group, WISE members explored feminism and used its implications for science education to transform our ways of seeing society, schools, and science classrooms. Without this opportunity to critically examine science and science education, I would have continued to perpetuate and reproduce the traditional view of science that privileges the masculine, reinforces existing power structures, and promotes a self-sustaining system of inequality (Mayberry, 1998). Once my doctorate was in hand, I became the secondary science methods and procedures instructor in the Teacher Education Program at the University of California, Santa Barbara. I began to strongly encourage the science student teachers I worked with to reexamine their own experiences and beliefs about science and what it means to be "scientific." Together, we critiqued how science is represented in society and in schools and struggled with what it means, in a state as diverse as California, for all students to acquire the knowledge and skills needed to be scientifically literate.

My work with science student teachers in the contexts of schools made visible the need for ongoing professional development after teacher preparation had ended. I saw that introducing student teachers to feminist approaches to science teaching and learning created conflicts with experienced teachers who privileged academic content and promoted replication of laboratory experiments. What happens to student teachers when they leave teacher education programs and become socialized into the culture of traditional secondary schools? I wondered. Are they willing and able to break traditions and push for innovative science curriculum and approaches that help students make personal connections to science and develop an awareness of the cultural, social, and political contexts within which science is practiced and used? I also know that research studies on professional development identify the power of teacher-teacher professional collegiality as a key to school success and effective school change (Griffin 1991). In particular, I believe interactive professional development is most powerful when caring and trustworthy peers are given the opportunity to learn from and with each other in a safe space, in particular, when women's talk and ways of knowing are valued and empowering relationships are cultivated (Cavazos 1994; Rogers 1993). I thought women science teachers, new to the profession, needed a safe, supportive learning community to talk about concerns, questions, and issues that arose as they balanced their personal and professional lives and negotiated the multiple roles and responsibilities they assume as women, scientists, and teachers.

Thus, in an effort to provide both professional support and opportunities for inquiry for women science teachers who graduated from the Teacher Education Program at UCSB, I organized the Women Educators of Science and Technology group in 1996. Our inquiry and research is centered around three questions. One, what are the major obstacles beginning and experienced teachers face when striving to be active, innovative members of the science teaching profession? Two, what kinds of support systems, including people, resources, and strategies, are essential for women science teachers who promote a feminist approach to the teaching and learning of science? Three, how can professional developers and science teachers work together to create inclusive curricula and instructional strategies that will help students find connections between their own lives and science concepts, observations, and experiences, as well as acquire an understanding of how science can be made accessible and useful for all?

As explained above, the women in the WEST group are former student teachers I taught and supervised during their fifth year credential/master's program. We began meeting as a group in the fall of 1996. The invitation to participate was offered to all women who graduated from UCSB in secondary science education since 1993; it continues to be offered to each cohort of women science teacher graduates. The initial group consisted of nine women, including myself as the participant-researcher. As of winter 2000, the group includes thirteen members, three of whom are first-year teachers. The women in WEST share similarities in terms of gender, science teaching credentials, educational philosophies focused on inclusion and equity, and beliefs that the teaching profession is enhanced and strengthened by women's involvement and leadership. Our differences, which include age, ethnicity, science content background, and personal and professional experiences, add to the variety of perspectives we bring to our work as teachers, scientists, and women.

The WEST group meets on a monthly basis during the academic year to explore current issues in science education, examine the support systems teachers have and need, and establish areas of curricular and instructional development relevant and fulfilling to all in the group. During the past three years, we have focused on crafting curricula that integrate content and scientific inquiry, provide all students with hands-on experiences, and offer multiple ways to be assessed. We recently developed a comprehensive "Aquarium Adventure" unit that emphasizes data collection and interpretation. We continue to work collaboratively to move beyond the stories we share about gender bias, inequitable learning opportunities, and the challenges of standards-based curricula toward changing our practices to include content and strategies that present a new image of science, scientists, and successful learning to students.

Julie Bianchini and Promoting Women and Scientific Literacy. Educators and scholars of science have described how the culture of science—its norms, values, shared meanings, and methods—creates many and varied obstacles for women and ethnic minorities as they move from elementary school classrooms to undergraduate science courses to professional laboratories (Brickhouse 1994; Harding 1993; Ginorio 1995;

Lemke 1990; Rosser 1991, 1995, 1997; Tobias 1990, 1992; Traweek 1988). Although many scholars have examined the cultural barriers women and ethnic minorities face in learning and doing science, few have attempted to use such critiques to design or research inclusive courses, especially at the college level (Rosser 1995). In response to this need, in 1997, science, science education, and women's studies faculty at ten universities across the United States embarked on a three-year curriculum and faculty development initiative designed to transform undergraduate science and women's studies courses. The project, Women and Scientific Literacy: Building Two-Way Streets, is coordinated by the AAC&U and funded by NSF. As a former science education faculty member at one of these ten universities, I became involved in that institution's initiative to make undergraduate science education more inclusive: the Promoting Women and Scientific Literacy project.

Goals for the Promoting Women and Scientific Literacy project established at this urban university exist at two levels. According to AAC&U (1996), the larger project's goal is to promote communication between science and women's studies departments: to infuse new scholarship about gender and science into undergraduate science, engineering, and mathematics education and to make science a more central component of women's studies courses. "The project takes up the challenge posed by biologist Anne Fausto-Sterling to break the cycle of reproducing 'a world in which science seems an illegitimate place for women and gender studies seems an inappropriate enterprise for scientists'" (1996, 2). The project members at the one university under study drew from AAC&U's stated goal to craft three primary objectives for their own initiative: one, to increase faculty awareness, sensitivity, and knowledge related to issues of gender, ethnicity, and the nature of science; two, to design and share pedagogical strategies to make science education more inclusive; and three, to promote the revision of courses to incorporate more inclusive science content and pedagogy. The local project expanded AAC&U's initial focus from issues of gender to that of gender and ethnicity; its purpose became to make both major and general education science courses more inclusive of the many women and ethnic minorities enrolled in them.

As with its goals, the Promoting Women and Scientific Literacy project includes two levels of participants: team members and project participants. The project team consists of ten members: six European American women (of which I am one), one African American woman, one Asian American woman, one man of mixed ethnicity, and one European American man. Team members span the disciplines of the biological sciences, women's studies, chemistry, science education, and psychology. Three are also chairs of departments. As a member of this team, my role is that of faculty development coordinator, project co-evaluator, and educational researcher. Project participants continue to grow in number and diversity of experiences. In the first year of its existence, during the 1997–1998 academic year, project participants included twelve science professors and one science educator; these professors were invited to join the project because they both teach introductory science courses and are considered open to issues of inclusion and instructional innovation. Six of these participants self-identified as European American women, four as European American men, one as an Asian American man, one as a man of mixed ethnicity; and one man simply as *Homo sapien*. In each successive year, both in 1998–1999 and 1999–2000, additional faculty members from across the sciences, mathematics, and engineering have been invited to participate in the project as well. Like the WEST group, participants in the Promoting Women and Scientific Literacy project are interested in transforming their practice toward the goal of including all students.

To help achieve the project's stated goals, beginning in August of 1997, the first of two years of faculty workshop sessions were developed and implemented. As stated above, the sessions were expected to inform scientists about issues related to gender and ethnicity in science and science education, as well as to support them in their efforts to make targeted undergraduate courses more inclusive. I organized these professional development sessions, in consultation with a women's studies professor and, to a lesser extent, with faculty participants; over time, primary responsibility for selecting topics and leading sessions shifted to the scientists involved in the project. In this first year of professional development seminars, topics ranged from the experiences of local women science students and scientists, to common ethnic and gender stereotypes related to science and science education, to past and present contributions of women in science and resources for locating them, to gender and racial bias in scientific research, to inclusive instructional and assessment strategies, to the mentoring of students from underrepresented groups. Most sessions were supplemented by scholarly readings pertaining to the topic at hand: journal articles from *Science Teacher*, *Journal of Research in Science Teaching*, and *Science Education*; selected passages from *Women of Science: Righting the Record* (Kass-Simon and Farnes 1990), *Teaching the Majority* (Rosser 1995), *They're Not Dumb: They're Different* (Tobias 1990), and *Warming the Climate for Women in Academic Science* (Ginorio 1995); and University pamphlets and brochures.

For many faculty participants, these readings served as introductions to the fields of science education and feminist science scholarship.

In year two of the Promoting Women and Scientific Literacy Project, additional professional development seminars were held; although both first and second year participants were invited, few first year members were able to attend. To develop greater ownership of the project and to build a stronger network of support for innovation, participants constituted five action research groups, each with a different focus of inquiry. One group began to explore issues related to the retention of Latinas in science; a second, to incorporate collaborative learning into instruction; a third, to develop active learning strategies for large lecture courses; a fourth, to investigate aspects of scientific writing; and a fifth, to initiate changes in a biology course for general education students and prospective elementary teachers. To more tightly integrate feminist science scholarship with science issues of particular relevance to women, members also developed two new undergraduate science and women's studies courses—Women and Science, and Science and Society—and extensively revised a third—Issues of Women's Health.

In the third year, project participants moved from attending professional development seminars toward working collectively to institutionalize project activities. Innovation and inquiry in the action research groups continues. In addition to the new and revised courses for undergraduates discussed above, a course for science teaching assistants has been established. Moreover, members have initiated conversations with the provost about ways to position gender and ethnic equity for faculty and students in the sciences as strategic priorities of the University—to integrate issues of equity and access in the sciences into the long-range vision of the institution as a whole.

Research Questions and Method

Research questions reflect both models of curricular and instructional transformation discussed in our Conceptual Framework and the goals and organization of our respective professional development projects. Answers are intended to inform other professional developers interested in supporting practicing science teachers toward the goal of inclusive science education. Questions include: Once teachers move beyond awareness of issues of gender and ethnicity in science, how do they attempt to integrate inclusive content and pedagogy into their everyday educational practices? What factors shape their ability to plan and implement a transformed curriculum? What successes and/or challenges do they encounter along the way? What can we as professional developers do to better support and sustain science practitioners' efforts at inclusion? How can we use our experiences to inform other professional development projects?

Data to fashion four case studies of teachers enacting curricular and instructional innovations were collected by the two authors in their respective projects using different means. Cavazos audiotaped monthly meetings held by the WEST group across four academic years, from 1996 to the present, as well as conducted a series of brief individual interviews with the two teachers selected for this study. The monthly meetings included four forms of conversation: dialogic (directed toward discovery and new understandings); conversation (a highly active and engaged form of talk about practice); narrative (stories which integrate intuition, experience, readings, and acquired knowledge); and talk (anecdotal stories of frustrations or celebrations). Cavazos examined all audiotapes of meetings and interviews, transcribing those excerpts she deemed relevant to the research questions posed in this paper. Bianchini, in turn, collected data from the Promoting Women and Scientific Literacy project with considerable assistance from her colleague, David Whitney. During the first year of the project, in 1997–1998, team members and participating scientists were individually interviewed prior to and after completion of the yearlong professional development seminar series. Both interviews were semi-structured in design and included approximately fifteen questions. Questions ranged from those about course organization and purpose, to types of pedagogical strategies employed, to views of gender and ethnic bias in science, to explanations of student learning. Each interview lasted approximately forty-five minutes, was recorded on audiotape, and was transcribed in full. (For additional data from these two projects, see also Bianchini, Cavazos, and Helms in press; and Bianchini, Whitney, Breton, and Hilton-Brown 1999.)

Four case studies were then constructed from qualitative analysis (see Erickson 1986; Hammersley and Atkinson 1983; Spradley 1980; Strauss 1987) of resulting transcripts. Our decision to develop these case studies was influenced by Lancy (1993), who described cases as the method of choice for studying interventions or innovations in educational settings. In keeping with the recommendations of Kreinberg and Lewis (1996), we approached the selection of possible cases with utmost respect for the expertise of participants and with an eye for diverse approaches to the equity challenge. We decided to strategically select cases (Hammersley and Atkinson 1983) of best practices, to examine participants who deliberately and successfully implemented culturally inclusive and gender equitable content and strategies. Sampling within each of these four cases (Hammersley and Atkinson 1983) was performed as well. Transcript contents were sorted into categories of cultural meaning, or domains, and then subcategories within each (Spradley 1980). Four domains of analysis were created: conceptions of inclusive science education, reasons for promoting inclusion, examples of inclusive educational practices, and challenges to implementation. Once initial cases were crafted, we asked participants to check them for validity; in other words, to inform our revisions, those studied were invited to critically react to our descriptions of their educational practices (Erickson 1986). Ultimately, these careful descriptions of practices instituted and challenges encountered by teachers in our professional development projects are expected to be of use to other science teachers and professional developers interested in inclusion—to be used by them to compare, contrast, and ultimately improve their own teaching of science (Elliott 1991).

Results: Successes and Challenges in Implementing Inclusive Science Content and Instruction

As stated in our introduction, our purpose is to identify lessons learned from our own efforts at professional development, to inform other teacher educators, science teachers, and university scientists interested in implementing an equitable and excellent science education for all students. To make visible the strengths and limitations of our efforts, we crafted case studies of four participants—two science teachers involved in Cavazos' WEST group and two scientists participating in Bianchini's Promoting Women and Scientific Literacy project. We presented the case of Denise at the beginning of our paper; those of Andrea, Elaine, and Cynthia are included below.

The Case of Andrea

Like Denise introduced above, Andrea is a junior high science teacher and member of the WEST group. A Hispanic woman, Andrea entered the teaching profession after working as a research scientist with NASA—after having the opportunity to be involved in the experimental side of the scientific enterprise. She became a teacher to share her love of science with students and to acquaint them with science outside the classroom setting. At present, Andrea is in her fourth year of teaching and works with many of the English Language Learners (ELL) at her site. She is also the MESA (Mathematics, Engineering, Science Achievement) coordinator for her school, an organization that encourages minority students to consider careers in math, science, and engineering.

In her teaching, Andrea focuses on the process of "doing science" and designs learning activities for students, activities that require critical thinking and problem solving, rather than extensive memorization of facts. For Andrea, inclusive science education means students have the opportunity to think through a problem or issue without worrying about finding the "right answer." Science becomes exclusive, Andrea explains, when "students focus on memorizing scientific facts, theories, and formulas that someone famous has validated and identified as important to learn. When science becomes the right and wrong answers, those students who do not have the right answers become frustrated and excluded. When all students gain an understanding of a science concept, everyone wins and feels included." Rather than excluding students by emphasizing right and wrong answers, Andrea continued, she attempts to include all students by creating opportunities for each to voice her or his opinion and to explore each others' ideas. "When we work together as a class to come up with a conclusion, we select from all of their ideas. Every student's input is valuable."

An example from Andrea's seventh grade earth science curriculum illustrates how she combines content and pedagogy to provide an inclusive learning experience for all of her students, especially for those who are struggling with the English language and/or with sophisticated science content. The following series of lessons examines the construction of the theory of continental drift.

> I begin this unit by having the students look at the globe and try to determine if any of the continents look like they could fit together like pieces of a puzzle. They are required to write down their ideas and they share their ideas with the class. Then, we cut out model pieces of the current continents and each student fits them together as they choose. They are not always sure of what is the best fit and everyone's looks different but that doesn't matter because we are concerned with them focusing on problem-solving not what is the right way to fit them together. At this point, I introduce them to Alfred Wegener, a German scientist who first introduced the idea of continental drift back in the early 1900's. I explain to the students that he did the same thing they just did because he was investigating his ideas about continental drift and if the continents were ever connected together as one large land mass. I want them to see that this is how scientists work—that science is about problem solving and not always about being objective and unbiased. It is about trial and error and scientists trying to come up with explanations that can be supported with evidence. We then talk about the idea of continental plates and how it is possible for continents to move across the surface of the

earth. I introduce fossil evidence at this point and we talk about glacier movement, coal deposits, and the locations of mountain ranges found on the continents. The students then draw in the mountain ranges on their continent pieces and color in areas of coal deposits. Now they are ready to match their continent puzzle pieces using the fossil evidence matching mountain ranges, coal deposits, and other fossil evidence to improve their guesses about how the pieces should fit together. I want the students to move beyond the puzzle shapes and see how evidence and problem solving are used in science. . . . We culminate this unit by watching a laser disk on Pangaea and students are now able to see millions of years of continental movement and consider how scientists used the evidence we have to develop the theory of continental drift.

In attempting to implement inclusive science content and instruction through problem-solving units, Andrea, like Denise, experiences her own set of challenges, challenges that Lynne, and her colleagues in the WEST group, attempt to help her address. One set of challenges identified by Andrea arises from her efforts to develop problem-solving units, like the one described above, that incorporate the contributions of many and varied scientists, especially women scientists and scientists from underrepresented ethnic groups. Andrea finds that textbooks and resource guides rarely include the kinds of information necessary to create activities that introduce students to the thoughts and actions of underrepresented scientists. Alternative sources of information, she notes, are difficult to identify and to obtain.

Andrea's second set of challenges involves her conception of what science is and what it means for students to learn scientific concepts. Andrea tries to patiently "unteach" students' misconceptions about the process of doing science—to demonstrate to students that both scientific discovery and science learning are messy and time consuming endeavors. Because students tend to focus on getting the "right answer," Andrea explains, they are often reluctant to exhibit creativity or to venture tentative explanations in the science classroom. Many students have acquired the belief that science is about right and wrong answers, not about problem solving and discovery as a result of their prior experiences in science classrooms. Andrea works to make clear to her students that "all the questions in science have not been answered and that they will be providing answers to scientific questions in the future as ideas change and new information is gathered."

A third set of challenges centers around issues of time. Andrea continues to struggle to slow down her instructional pace so that students have adequate time to wrestle with and complete the complex problem-solving activities she designs. She also thinks that "the [recently adopted California science] standards do not emphasize problem solving or encourage students to discover and learn for themselves" and worries that she does not have enough time to teach students both standards-based content and problem-solving strategies. She views "the new California standards [as] forc[ing] teachers to emphasize content over process to ensure that students will excel on district-adopted standardized tests." Thus, Andrea needs assistance from other WEST members to carefully examine and critically reflect on her chosen approach to inclusive science education in light of the current emphasis on standards-based reform. Is her problem-solving, process-based approach truly a more inclusive way for teachers to teach and students to learn science? Does it provide adequate opportunities for all students in her classroom to acquire the scientific knowledge, skills, and habits of mind needed to become scientifically literate? Can she justify the time required for this kind of teaching given California's current emphasis on student achievement and teacher accountability as measured by standardized test scores?

The Case of Cynthia
Cynthia is an Asian American biologist and team member of the Promoting Women and Scientific Literacy project. In partial response to project goals and expectations, Cynthia has worked to revise her general education, interdisciplinary course, Human Immunology: In Self-Defense. Course content and instruction have been refashioned to reflect three principles of inclusion. For example, Cynthia now foregrounds the notion of context in her presentation of scientific information. Cynthia explains, "I think a faculty member who . . . doesn't provide any kind of a context, [who doesn't] explain how this material is applicable or significant is not going to get the information across" to students. "Because males have been the principal researchers and have written most of the [science] textbooks, science has been interpreted from a male viewpoint," she continues. Science students need to understand that "it makes a difference in interpretation where you come from and in what kind of context you place information." Cynthia has also revised her immunology course to better address the diverse needs and interests of her students. Many science faculty, Cynthia notes, think "everybody's the same" and thus believe "treating everybody the same" is treating all equally. Cynthia, however, sees such so-called equitable treatment as a barrier to female and ethnic minority students who have different learning styles and different ways of interacting. Because "women do learn a little differently than men," Cynthia elaborates, she makes an extra effort to include group work and to ensure that

typically reserved Chicana and Vietnamese women participate. Finally, in restructuring her course, Cynthia has begun focusing on the big ideas of science rather than its numerous and detailed facts. "I've decided it is not as important to give students all the minutia. If they can get the basics and if they can appreciate [the big ideas], then they will remember something about the topic" in years to come.

Thus, in her revised Human Immunology course, Cynthia attempts to implement innovative instructional strategies and gender equitable, culturally inclusive content. "In order to promote better participation and more interaction" among students in her course, for example, she created and regularly implements "Fun Fridays." Fun Fridays bring "in more of the ethical kinds of considerations or some of the really current things that are happening" related to human diseases. "Every Friday students are randomly divided into groups. . . . During [the ensuing small group] discussions, students compare notes, ask questions about lecture and text material covered that week, and discuss . . . the significance to society, ethics, and applicability of the information in [an assigned] article." Cynthia also requires each group of students to complete one "panel presentation" sometime during the course. "By having an oral panel presentation, students practice oral communication skills and share with other students what they have learned. . . . Students enthusiastically participate and bring [in] not only factual information, but ethical issues and demonstration materials" as well. Cynthia's third interactive pedagogical technique is the posing of open-ended questions and the elicitation of participation of all students in her class: "Previously I had used primarily [questions that required] one word, right or wrong answers. I now try to pose more open-ended questions, encourage other students to join in and add further answers, and most importantly, wait longer for answers. . . . I also am much more aware about giving equal time to male and female students and encouraging female participation."

Finally, Cynthia has incorporated "concept mapping as a review technique for midterms." Students are divided into groups; given a different list of key concepts, colored post-its, a transparency sheet, and pen; and then asked to present their concept maps once complete. Cynthia thinks this strategy beneficial: "Students have to discuss with each other what these terms mean and how they are related. By presenting their results, they [also] help review their peers."

In addition to innovative instructional strategies, Cynthia has attempted to infuse new content into her course to spark the interests and raise the awareness of her diverse students. She presents the work of scientists from underrepresented groups and the contributions to science from other cultures in her lectures: "I include at least a couple of women [scientists and] . . . talk about the Chinese and small pox vaccine. . . . I [also] do an introductory section . . . at the beginning [of the course] because I've always believed that you need a context in which to place what is known with . . . what's happened historically." To complement these discussions of contributions, Cynthia also highlights the importance of diverse perspectives on scientific practice. She discusses with students recent trends related to AIDS and women in the United States: "I try to point out where women have had an influence, because otherwise medicine is usually from a male point of view. . . . We talk about AIDS as one of our last topics and so I make it a point to make sure the students realize that for many years, they didn't have a definition of AIDS that applied to women. . . . It's only been in the last five years that they've included vaginal yeast infections, chronic vaginal yeast infections, as a marker for an AIDS infected female."

As with Denise and Andrea, Cynthia recognizes the value of professional development opportunities afforded her. She believes she has benefited from attending the professional development seminars as part of her participation in the Promoting Women and Scientific Literacy project. She finds discussions with other faculty involved in the project—an all too rare occurrence—particularly useful. "One of the most helpful aspects of the Promoting Women and Scientific Literacy grant has been exposure to and sharing of philosophy and pedagogical techniques used by other faculty," she explains. "Otherwise we tend to operate in isolation, only doing what is most comfortable and convenient to us."

Still, Cynthia identifies two persistent challenges encountered in continuing efforts to refine and improve her course. One challenge involves the inclusion of content relevant to all students and its presentation in an effective fashion. Science courses, she explains, "are very information bound." In her immunology course, she must "cover normal structure and function of cells and various parts of the body" before she "can talk about the disease aspects. . . . If you just jump into the disease aspects, it doesn't make sense." The coverage of basic facts, human physiology, and diseases, however, leaves her little time "to include the research that people have done," "to include a lot about the contributions of women and minorities" to science. The use of innovative instructional strategies requires an additional "sacrifice of informational content"—both "a trade-off" of less content covered and less control over the quality of that coverage. Cynthia says, "I was very pleased that I got more student participation [by implementing Fun Fridays and panel presentations]. It means, however, you have to give up some of your lecture time. The material, even though some of it will get covered by the panel presentations students are giving, may not all be covered and may not be covered very well either."

Cynthia also continues to search for ways to help undergraduates adapt to her different set of classroom norms and expectations. She has found effectively managing students during group work a challenge; she has been unable "to figure out a way to assure that everybody does their fair share." Some students, she explains, fail to "read the assigned article and contribute to the discussion." Other students are reluctant to engage in group conversations because they "don't have it in their culture to be outspoken and more participatory." Still other students complain that "not everybody's doing all they're supposed to do in that group." Helping students see the benefits of moving from passive note taker to active participant also has not been easy. Because students must actively participate during Fun Fridays and panel presentations, Cynthia notes, they "have to do a lot more work. They aren't pleased about it because it means that they have to go out there and do the research and look up the articles and put [ideas] together."

The Case of Elaine

Like Cynthia, Elaine is a member of the Promoting Women and Scientific Literacy project; she is the subject of our fourth and final case study. A European American geologist, Elaine has worked for years to make her large introductory geology course more inclusive of women, ethnic minorities, students with disabilities, and economically disadvantaged students. Her primary objective is to help as many students as possible succeed in her course: "I try to do whatever works. . . . The dominant learning strategy is to have the students perform well on exams and to facilitate that in however I may." A second aim is "to assist the students in gaining an understanding and appreciation for the earth as a system, as illustrated by key [geological] concepts." Her third "major goal is to attend to the aims of the general education program in skill building." Because the introductory geology course is expected to satisfy general education requirements, Elaine provides multiple opportunities for students to practice and improve their oral and written communication skills.

Elaine has implemented a wide array of inclusive instructional strategies in her introductory geology course—too many to describe in this short case. To "really get people out of the lecture mode," for example, Elaine includes a weekly "in-class writing assignment that often involves interactions with people in adjacent seats or formation of small teams of people to address questions." She integrates technology into her course as well: She assigns an internet project, requires students to use the textbook's CD, and is "in the process of building internet modules that facilitate asynchronous learning." Although sensitive to "classist assumptions . . . [that] students all have equal access to technology," she thinks it imperative each student gains experience using technology and believes technology can "enhance learning for people that are very visually oriented." Elaine also uses "personalized strategies" to better meet the needs of physically and learning disabled students. These include placing her lecture notes in the library, extending personal invitations to attend office hours, and holding reviews before exams. She explains, "I have been very happy to work with Disabled Students Services and to try and make as many options open for people who don't have what we consider standard abilities to process information. . . . I've been very interested in learning more about how I can proactively identify [learning disabled] students who haven't actually been tested yet." Finally, Elaine has expanded her assessment repertoire to include writing assignments and class participation as well as exams: "More and more I'm determining their grades following a loose portfolio style assessment, where I'm looking at their grades on exams, their performance on the paper, [and] their participation in class. . . . I keep trying to push the envelope to decrease the amount that relies on exams simply because my major goal is to ascertain what people know."

Elaine has focused more on expanding the kinds of instructional strategies she employs than on revising the subject matter she discusses in her course; still, several innovative pedagogical techniques carry with them more female friendly, culturally inclusive science content. The last time she taught introductory geology, for example, students were given extra credit for attending the presentation of a visiting woman scientist and writing about their experience. "I think that visit in particular was very important for a lot of the students in the class," Elaine notes. "I think that a real positive image of a woman as a geologist practicing in the field was obtained." Elaine also assigned a "five page paper [that] had to do with interviewing individuals who had lived through a natural hazard such as an earthquake, volcanic eruption, a tornado and then also providing the scientific background to that hazard, why did it occur, what happened, what were the effects." The paper assignment proved inclusive in respect to the hazard investigated: "It was an activity intended to be inclusive of underrepresented groups because many people chose hazards in countries other than the United States. . . . Most of the people that were chosen to be interviewed tended to be from the Philippines or from Mexico. So when I would read the papers . . . it was obvious that there was a lot of identification with, linking geology with experiences in home countries. . . . So, that was something that was very subtle but . . . if I'd said to all of them, "Write a paper on the eruption of Mount St. Helens," . . . then it would have not achieved that objective."

Like the other science teachers and scientists presented before her, Elaine needs additional support from colleagues and professional developers to think through and attempt to resolve persistent challenges faced in her efforts to implement a science education inclusive of all students. Elaine, for example, finds adopting gender equitable and culturally friendly instructional strategies easier and more expedient than modifying course content. She has been unsuccessful in locating concrete examples of inclusive science subject matter that move beyond accounts from the history of science. Besides including greater discussion of the contributions of women and ethnic minority scientists as well as the sciences from other cultures, Elaine continues to search for ways the substance of geology can be transformed to include all students.

> The content [of geology] in and of itself does not necessarily lend itself to that type of inclusivity. . . . See the styles, we can do a lot more work on the styles of learning and the styles of assessment to my mind. It's much easier for me to work on my own approach to that and learn from other people than to say, "Well, a sedimentary rock." I mean a sedimentary rock is a rock; it's not a human. . . . It's gender neutral and the same thing with natural processes such as sedimentation and volcanism. So then you become very reliant on anecdotal examples in which individuals from formerly underrepresented groups are figured prominently in that research. I definitely try and do that. I'm just really interested in how you could possibly do more given the content we have. . . . I've even ordered books about how to find more ways of doing this. We don't get a lot of guidance in literature right now.

Elaine also finds including discussions of the contributions of underrepresented groups challenging in and of its own right. The introductory geology curriculum as currently envisioned and implemented across the country, Elaine explains, provides little room for the discussion of people and their contributions. She stated, "That's a conscious decision [our department made] based on the amount of material that we should be covering to be in parity with other departments [at other institutions] offering the same class. Now that may change in the future . . . but at this time, we're not out there at the forefront saying, We're. . . going to focus only on the people and have the students learn from that vantage point."

Introductory geology textbooks, Elaine continues, provide little assistance as well: Few include discussions of white male scientists let alone the work of women and ethnic minorities. "Geology is a culture that began in the British Isles and so all the textbooks focus on a very narrowly defined group of people," Elaine clarifies. Textbooks only spend "about 1% of the text talking about people. [They talk] about observations people have made, detached from the people that have made them."

Finally, like Cynthia, Elaine finds her undergraduate students do not necessarily recognize or sufficiently appreciate her efforts to make science more inclusive. For example, although students' final grades improved the last time she taught the course, students had to "work harder so their satisfaction with the course as measured by student evaluations went down." "There are a lot of students," Elaine adds, who are "not interested in group work. . . . They're not interested in hands on. They're interested in taking the materials, drawing in what's happening in a standard lecture, internalizing it, and then coming back." As a result, Elaine continues to question how best to ensure instructors do not lose either group of students, those students that thrive under traditional instruction and those that respond best to innovative approaches.

What Can We Learn from Practicing Science Teachers and Scientists?

We draw from our experiences in professional development, from reflections on our conceptual frame, and from analysis of the successes and persistent challenges of our participants to offer a series of recommendations to other professional developers, science teachers, and scientists interested in promoting a science education inclusive of all students. We begin by reminding readers that the four teachers presented in this chapter each constructed individual visions of inclusive science teaching and learning responsive to their own interests and science backgrounds, students' needs and experiences, and demands imposed by institutions of public education. The need for teachers to create their own vision of inclusive science education was underscored in our conceptual framework (see Kreinberg and Lewis 1996; and Middlecamp and Subramaniam 1999). Denise, for example, strove to integrate subject matter across disciplines and to use strategies that recognized multiple ways of knowing to help all her students understand and make personal connections to science concepts. Her integrated, non-linear approach reflected her training in environmental sciences and way of thinking about the interconnectedness of scientific concepts. To attend to the needs of her students, especially those with physical or learning disabilities, Elaine worked closely with Disabled Students Services to provide multiple opportunities for students to acquire the content knowledge and technological skills needed by practicing geologists and scientifically literate citizens. Elaine also chose to focus on implementing a wide array of inclusive teaching strategies to accommodate students' different styles of learning and their different preferences for forms of assessment. Finally, given California's current emphasis on standards-based instruction and teacher accountability, Andrea wrestled with institutional constraints: She struggled to balance district demands to increase student test scores with her desire to provide students opportunities to develop their critical thinking and problems solving skills. As made clear by looking across the cases of Denise, Elaine, and Andrea, then, neither generic questions nor uniform solutions to equity issues exist; instructors, immersed in their own personal and professional contexts, encounter challenges and devise practices never envisioned by the professional developers with whom they work.

A second lesson we learned from participants in our professional development efforts is that the identification and integration of inclusive science content can be more difficult than implementation of inclusive instructional strategies. Denise, for example, was unable to collaborate with other science teachers in her department because few had the knowledge of disciplines outside of science required to create and implement integrated units. Andrea had difficulty locating information about the lives and work of women and ethnic minority scientists. Elaine experienced frustration in trying to locate concrete examples of inclusive content that extended beyond the contributions of members from underrepresented groups. In addition, Andrea, Elaine, and Cynthia each described pressures they felt to cover content broadly rather than deeply, to discuss traditional science content rather than topics of greater relevance to students' lives. Thus, to help address tensions related to the kinds and amount of content to cover, we urge professional developers to provide opportunities for science practitioners to explore scholarship related to the history and nature of science; examine intersections between science topics and other disciplines; identify resources, institutes, and organizations from which to acquire necessary information; and work together to change conceptions of science content from mere facts and concepts to ones that include historical accounts, real-life applications, and interdisciplinary investigations. Because innovative strategies are necessary but not sufficient to render science education inclusive (Mayberry 1998; Middlecamp and Subramaniam 1999), science teachers and scientists need time to locate, talk about, and come to understand nontraditional and revolutionary science subject matter.

Our third recommendation is that professional developers work with teams of teachers or scientists at a given site rather than individuals from diverse locals to effect change. As noted by both Loucks-Horsley et al. (1998) and Kreinberg and Lewis (1996), such collaboration is of direct benefit to practitioners: Teachers can provide each other needed support and diverse perspectives grounded in common experience. As Cynthia expressed in her interview, one of the most valuable aspects of the Promoting Women and Scientific Literacy project was the opportunity to exchange ideas and compare practices with other colleagues—opportunities rarely afforded instructors in the current organization of schools and universities. Equally important, such collaboration can aid teachers and scientists in their efforts to help students more readily adapt to nontraditional course requirements and instructional activities. As noted by each participant in our case studies, students' resistance to innovation made efforts to implement inclusive approaches more difficult. Andrea's students, for example, were reluctant to think creatively and pose possible solutions. Elaine's students thought the in-class activities and writing assignments for her introductory geology course an inordinate amount of work. Professional developers, then, must be cognizant that students experience schooling as a comprehensive endeavor rather than as discrete and isolated classes. They must work to bring teachers within school settings together so that expectations for and activities required of all students are revised across multiple courses and years of instruction—so that both teachers and students can support each other in teaching and learning science inclusive of all.

Fourth, from our examination of participants' experiences, we echo Loucks-Horsley et al.'s (1998) argument that a strong professional culture is essential for initiating changes in content norms and pedagogical practices within science education (see also Thomas, Wineburg, Grossman, Myhre, and Woolworth 1998). A professional community can only be formed when science educators work with colleagues as members of a long-term initiative to identify, explore, and adapt strategies for integrating inclusive content and pedagogy into their educational practices. As made clear by the models of Rosser (1991, 1995, 1997), Banks (1999), and Nieto (1996) discussed in our Conceptual Framework and the four case studies of participants presented in our Results section, it takes an inordinate amount of time—years rather than days or months—for science teachers and scientists to re-examine their own experiences, principles, and practices about science and science education and to implement coherent and consistent strategies to combat the inequalities that currently exist at all levels in our educational system. Educators must expend years of effort to break away from traditional curriculum and instructional practices; acquire a philosophy of teaching and learning that affirms and supports all students; and most importantly, work to resolve the myriad challenges that arise in providing all students access to an excellent and equitable science education. We believe that with adequate time and within supportive learning communities, science teachers and scientists can acquire the skills needed to implement a student-centered approach to instruction, obtain the necessary resources to move toward presenting inclusive content, and develop effective solutions to address the difficulties that inevitably surface as they attempt to transform both the content and pedagogy of their practice.

In closing, we remind professional developers that science teachers and scientists need not only *time* to thrive within professional communities, but also the *freedom* to enact decisions they perceive of greatest relevance to themselves and their own students. We thus recommend participation in professional development projects centered around inclusion to be voluntary: All participants in our two projects made a conscious and concerted decision to take up the equity challenge. Without this initial internal motivation, it is unlikely educators will take the time or exert the effort needed to transform their courses. For professional development opportunities to succeed, we also suggest professional developers view teachers as equal, competent partners in the equity effort and recognize that they have the knowledge, experience, and ability to transform traditional content and pedagogy into what is necessary and valuable for the particular group of students they serve. Roland Barth provides valuable advice to all involved in inclusive science education reform efforts: "Changes in schools may be initiated from without, but the most important and most lasting changes will come from within. Schools [and the teachers within them] are capable of improving themselves" (1990, 159).

Bianchini's study was funded, in part, by a Spencer Small Grant; the data presented, statements made, and views expressed, however, are solely the responsibility of the authors.

References

American Association for the Advancement of Science (AAAS). 1990. *Science for all Americans: A Project 2061 report on literacy goals in science, mathematics, and technology.* New York: Oxford University Press.

———. 1993. *Benchmarks for science literacy.* New York: Oxford University Press.

Association of American Colleges and Universities. 1996. *Women and scientific literacy: Building two-way streets. Application form.* Washington, DC: Author.

Atwater, M. M. 1996. Social constructivism: Infusion into the multicultural science education research agenda. *Journal of Research in Science Teaching, 33*(8): 821–837.

Banks, J. A. 1995. The historical reconstruction of knowledge about race: Implications for transformative teaching. *Educational Researcher, 24*(2), 15–25.

Banks, J. A. 1999. *An introduction to multicultural education* (2nd ed.). Boston, MA: Allyn and Bacon.

Barth, R. 1990. *Improving schools from within: Teachers, parents, and principals can make the difference.* San Francisco, CA: Jossey-Bass.

Barton, A. C. 1998. *Feminist science education.* New York: Teachers College Press.

Bianchini, J. A., Cavazos, L. M., and Helms, J. V. in press. From professional lives to inclusive practice: Science educators' views of gender, ethnicity, and science. *Journal of Research in Science Teaching.*

Bianchini, J. A., Whitney, D. J., Breton, T. D., and Hilton-Brown, B. A. 1999. Toward inclusive science education: University scientists' views of students, instructional practices, and the nature of science. Unpublished manuscript.

Brickhouse, N. W. 1994. Bringing in the outsiders: Reshaping the sciences of the future. Journal of Curriculum Studies, 26(4): 401–416.

California State Board of Education. 1998. Science content standards grades K-12 (pre-publication version). Sacramento, CA: Author. (See also.)

Cavazos, L. M. 1994. *A search for missing voices: A narrative inquiry into the lives of women science teachers.* Unpublished doctoral dissertation, Michigan State University, Lansing, MI.

Clandinin, D. J., and Connelly, F. M. 1990. Narrative and story in practice and research. In D. Schon (Ed.), *The reflective turn: Case studies of reflective practice.* New York: Teachers College Press.

Elliott, J. 1991. *Action research for educational change.* Philadelphia, PA: Open University Press.

Erickson, F. 1986. Qualitative methods in research on teaching. In M. C. Wittrock (Ed.), *Handbook of research on teaching* (3rd ed.) (pp. 119-161). New York: Macmillan.

Gardner, H. 1993. *Frames of mind: The theory of multiple intelligences* (10th anniversary ed.). New York: Basic Books.

Gaskell, J., and Hildebrand, G. 1996. Teaching individuals in a gendered world. *Reflect, 2*(2): 36-42.

Ginorio, A. B. 1995. *Warming the climate for women in academic science.* Washington, DC: Association of American Colleges and Universities.

Griffin, G. 1991. Interactive staff development: Using what we know. In A. Lieberman and A. Miller (Eds.), *Staff development for education in the 90s: New demands, new realities, new perspectives.* New York: Teachers College Press.

Hammersley, M., and Atkinson, P. 1983. *Ethnography principles in practice.* New York: Routledge.

Harding, S. 1993. *The racial economy of science.* Bloomington, IN: Indiana University Press.

Hurd, P. D. 1998. Scientific literacy: New minds for a changing world. *Science Education, 82*(3): 407-416.

Kass-Simon, G., and Farnes, P. (Eds.). 1990. *Women of science: Righting the record.* Bloomington, IN: Indiana University Press.

Kreinberg, N. and Lewis, S. 1996. The politics and practice of equity: Experiences from both sides of the Pacific. In L. H. Parker,

L. J. Rennie, and B. J. Fraser (Eds.), *Gender, science, and mathematics: Shortening the shadow* (pp. 177–202). Dordrecht, the Netherlands: Kluwer Academic Publishers.

Lancy, D. F. 1993. *Qualitative research in education: An introduction to the major traditions.* New York: Longman.

Lemke, J. L. 1990. *Talking science.* Norwood, NJ: Ablex Publishing

Loucks-Horsley, S., Hewson, P. W., Love, N., and Stiles, K. E. 1998. *Designing professional development for teachers of science and mathematics.* Thousand Oaks, CA: Corwin Press.

Mayberry, M. 1998. Reproductive and resistant pedagogies: The comparative roles of collaborative learning and feminist pedagogy in science education. *Journal of Research in Science Teaching, 35*(4): 443–459.

Mayberry, M., and Rees, M. N. 1999. Feminist pedagogy, interdisciplinary praxis, and science education. In M. Mayberry and E. C. Rose (Eds.), *Meeting the challenge: Innovative feminist pedagogies in action* (pp. 193–214). New York: Routledge.

Mayberry, M., and Rose, E. C. 1999. *Meeting the challenge: Innovative feminist pedagogies in action.* New York: Routledge.

McCormick, T. 1994. *Creating the non-sexist classroom.* New York: Teachers College Press.

Middlecamp, C. H., and Baldwin, O. 1995. The Native American Indian student in the science classroom: Cultural clash or match? *Proceedings from the Third International History, Philosophy, and Science Teaching Conference* (Vol. 2, pp. 776–787). Minneapolis, MN.

Middlecamp, C. H., and Subramaniam, B. 1999. What is feminist pedagogy? Useful ideas for teaching chemistry. *Journal of Chemical Education, 76*(4): 520–525.

Muller, C. B., and Pavone, M. L. 1998. The women and science project at Dartmouth: One campus model for support and systemic change. In A. M. Pattatucci (Ed.), *Women in science: Meeting career challenges* (pp. 247–265). Thousand Oaks, CA: Sage Publications.

National Research Council (NRC). 1996. *National science education standards.* Washington, DC: National Academy Press.

National Science Foundation (NSF). 1996. *Shaping the future: New expectations for undergraduate education in science, mathematics, engineering, and technology.* (A report on its Review of Undergraduate Education by the Advisory Committee to the Directorate for Education and Human Resources, NSF Publication No. 96-139). Arlington, VA: NSF.

National Science Teachers Association (NSTA). 1992. *Scope, sequence, and coordination of secondary school science. Volume I. The content core.* Washington, DC: NSTA.

Nieto, S. 1996. *Affirming diversity: The sociopolitical context of multicultural education* (2nd ed.). White Plains, NY: Longman.

Nieto, S. 1999. *The light in their eyes: Creating multicultural learning communities.* New York: Teachers College Press.

Price, J. N., and Ball, D. L. 1998. Challenges of liberatory pedagogy in mathematics and teacher education. *Theory into Practice,* 37(4): 256–264.

Rodriguez, A. J. 1998. Strategies for counterresistance: Toward sociotransformative constructivism and learning to teach science for diversity and understanding. *Journal of Research in Science Teaching,* 35(6): 589–622.

Rogers, A. G. 1993. Voice, play, and a practice of ordinary courage in girls and women's lives. *Harvard Educational Review,* 63(3): 265–295.

Rosser, S. V. 1991. *Female friendly science: Applying women's studies methods and theories to attract students.* New York: Teachers College Press.

Rosser, S. V. (Ed.). 1995. *Teaching the majority: Breaking the gender barrier in science, mathematics and engineering.* New York: Teachers College Press.

Rosser, S. V. 1997. *Re-engineering female friendly science.* New York: Teachers College Press.

Sanders, J., Campbell, P. B., and Steinbrueck, K. 1997. One project, many strategies: Making preservice teacher education more equitable. *Journal of Women and Minorities in Science and Engineering,* 3: 225–243.

Spradley, J. P. 1980. *Participant observation*. Fort Worth, TX: Harcourt Brace Jovanovich College.

Strauss, A. L. 1987. *Qualitative analysis for social scientists*. Cambridge, England: Cambridge University Press.

Sylwester, R. A. 1995. *Celebration of neurons: An educator's guide to the human brain*. Alexandria, VA: Association for Supervision and Curriculum Development.

Thomas, F., Wineburg, S., Grossman, P., Myhre, O., and Woolworth, S. 1998. In the company of colleagues: An interim report on the development of a community of teacher learners. *Teaching and Teacher Education*, 14(1): 21–32.

Tobias, S. 1990. *They're not dumb, they're different*. Tucson, AZ: Research Corporation.

Tobias, S. 1992. *Revitalizing undergraduate science: Why some things work and most don't*. Tucson, AZ: Research Corporation.

Traweek, S. 1988. *Beamtimes and lifetimes: the world of high energy physics*. Cambridge, MA: Harvard University Press.

Notes

1 We use pseudonyms for our participants.

2 Catherine Middlecamp and Banu Subramaniam both participate in the American Association of Colleges and Universities' Women and Scientific Literacy: Building Two-Way Streets, the same initiative of which Bianchini's Promoting Women and Scientific Literacy project is a part. Maralee Mayberry, whose ideas are discussed in a later paragraph, is also connected to this national initiative.

Chapter 11

Feminisms, Sacred Stories, and Multiple Voices

Sharon Parsons

Since 1991 I have undertaken an autobiographical inquiry into my practice as a science educator. My autobiography has been influenced by various positions within feminist theory (Parsons 1999). In this chapter I will address two of the challenges which have informed my research. My first challenge was the need to deal with the "sacred stories" of university practice, stories that contribute to maintaining the established power-knowledge relationships from multiple feminist perspectives. My second challenge was the need to deal with the sacred stories of theorized feminism that have informed my ongoing analysis.

Throughout my analysis I have become increasing aware of the wide range of positions within feminist theory and this has influenced my interpretation. While earlier in my analysis one theoretical position appeared to be the best interpretative perspective to bring to my practice, I now feel that a pluralistic perspective influenced by various "feminisms" (Luke and Gore 1992) is my preferred position. In arriving at this position I have also learned how to be less influenced by the alluring power of theorized feminism.

The Use of Autobiographical Narrative

My present assumption is that through an ongoing autobiographical study I can better understand some aspects of my practice from a "feminisms perspective" (Luke and Gore 1992), and at the same time take action to improve it. Some educators have criticized autobiography for being undertheorized. The theoretical underpinnings of autobiography however have been well reported in the literature, as evidenced by the work of Pinar (1988), and others (Brookes 1992; Bullough and Gitlin 1995; Connelly and Clandinin 1988, 1991, 1993; Cortazzi 1993; Davis 1996; Graham 1991; Pinar et al. 1995).

Given that feminist theoretical perspectives have informed my research, a feminist methodology and epistemology particularly suit an analysis of my actions as a science educator (Harding 1986, 1987; Hollingsworth 1994; Lather 1993; Middleton 1993; Noddings 1984; Robeck, Hepburn, and Gaskell 1995; Roberts 1981). In the analysis of my practice I have employed narrative inquiry (Connelly and Clandinin 1988, 1991, 1992, 1993; Cortazzi 1993). Connelly and Clandinin describe narrative inquiry as "inquiry into narrative" where the narrative is both the phenomena and the method, with the phenomena being the story and the inquiry the narrative (1991). The constant narrative inquiry into my practice since 1991 has been a useful method for me to reflect on my practice as a science educator (Parsons 1992, 1993, 1996, 1997, 1999; Parsons and Matson 1995; Parsons, Matson, and Rohan 1995).

My use of narrative inquiry to make sense of my practice is similar to Middleton's (1993) life-history approach, and to the use of autobiography by Brookes (1992) to construct a feminist pedagogy. Through my ongoing inquiry I have struggled to bring a feminist pedagogy to my practice as a science educator. This struggle, however, has been informed by my evolving understanding of feminist theory as I researched my practice.

I also see similarities between my work and what St. Pierre (1997) describes as nomadic inquiry. In St. Pierre's ethnographic study of white southern women in her hometown she uses "writing as a method of inquiry" to think differently about her fieldwork. Like St. Pierre, my writing/rewriting has allowed me to work out spaces (mental space, space of text, and the space of theory) within my practice. The difference is that I have primarily been interested in exploring my own space rather than the spaces of others.

I have found the telling/retelling of stories about my practice (as a feminist teacher and researcher) to be a powerful experience. Storytelling has allowed me to continually view my teaching as learning. Throughout the autobiographical process I have found storytelling to be both theorizing and transformative (Barton 1998). I have also found that the telling of my stories has created a tension for me between building theory through stories and using theories created by others to make sense of my practice. By using feminist theories created by others, at times, I have fallen into the entrapment of sacred stories of theorized feminism. In the next section, I will review the feminist theoretical positions that have influenced the interpretation of my stories.

Reviewing My Practice from Multiple Feminist Perspectives

In an earlier analysis (Parsons 1999) there were two broad theoretical perspectives which informed the analysis of my practice as a science educator: (1) an historical feminist perspective, and (2) a pluralistic feminisms perspective (Luke and Gore 1992) which has been informed by postmodern/poststructural and post-critical theoretical positions. Using these two theoretical perspectives as critical theorems (Rohan 1996), I completed an analysis of my practice at different points in time.

A Historical Feminist Perspective

In the beginning the history of feminist theory provided a basis for my analysis (Parsons 1993). In particular, I found the categories constructed by Kristeva (1982) to describe the history of feminist theory to be useful. She describes feminism as being first-, second-, and third-generation. In the first-generation, women seek equality with men, the typical liberal position. Liberal feminism is mainly concerned with unfair employment practices. In science, this is manifested as a political criticism in favor of equal opportunity for the sexes in science careers; it does not attempt to question the androcentric biases within science, or science careers. In the second-generation, women embrace their own special qualities and reject uncritical assimilation into the male world; the emphasis here is on moving the best female qualities into the public world. By the third-generation, women critique what they sought and accomplished in the first two cycles. Then they use that criticism to seek further solutions to old and new questions. The view of feminism presented by Kristeva is not merely descriptive but one of continuous evolution. Based on Kristeva's historical perspective, one could conjecture that feminism would continue to evolve and move beyond a third-generation.

Another label that has been used to describe some feminist theory is radical feminism. Radical feminism is important from a historical perspective because second- and third-generation feminist perspectives have been influenced by it. Radical feminism argues that, in the case of science, scientific ideologies and philosophies are based on androcentric foundations. This has lead to a masculine way of viewing science, one which in most cases also means Eurocentric science, a predominately white Anglo-

Saxon male perspective (Harding 1986; Keller 1985). A radical feminist perspective questions the nature of a scientific ideology based on androcentric foundations which permeate the social structure of science, its applications, and its methodologies (Keller 1985).

Feminisms: A Pluralistic Feminist Perspective Informed by Postmodern/Poststructual and Post-critical Theory

Much of postmodern/poststructural feminist literature to date has emerged out of liberal and radical feminist traditions (Gore 1992). For my purposes I will use postmodernism as a more general term to include poststructuralism, realizing that some feminist writers, particularly in Australia, Britain, and Canada, tend to use the term poststructuralism. Central to postmodernist theory is an anti-foundational epistemology which rejects foundational truths in disciplinary knowledge and the unitary rationalist subject as foundational to all knowledge (Luke and Gore 1992; Nicholson 1990). Postmodern feminist theory therefore rejects the universal subject, theories of the subject and others, and the social structures and theories that contain the masculine and feminine subject.

Nicholson (1990) adds to the above discussion by posing a significant question: Who are the women we are talking about? We are not homogeneous; we differ by class, race, culture, and sexual orientation. These differences suggest that multiple feminist voices be heard, rather than a dominant feminist voice (white, Western, and academic). While the postmodern position is not a perfect match for feminist theory, it is, nevertheless, a workable set of identity papers, a set that allows feminists to communicate with a broader audience.

Giroux (1992) notes that two of the most important challenges to modernism have come from postmodernism and feminism. Given the ongoing critique of modernism in science education this last decade, postmodern feminism has provided further insight into what a transformed science education might look like (Harding 1986, 1987). Gore's (1992) interpretation of a postmodern agenda, however, differs from Giroux's in that Gore wishes to apply regimes of truth to more than one society (the dominant society). Gore also acknowledges that feminisms may have their own power-knowledge nexuses which, in particular historical moments, will operate in ways that are oppressive and repressive to people who are within and/or outside that society. This is well illustrated in the anger that many women of color have expressed at the alienation and marginalization from primarily white, middle-class feminism in the academy. Drawing upon the notion of "regimes of truth," Gore reminds us that we need to consider the potential dangers and normalizing tendencies of all discourses, including those that aim to liberate others. Luke and Gore (1992) note that postmodernism/poststructuralism is problematic in that its theoretical tenets have been helpful only to the extent that they fit with feminist work. They prefer instead to label feminisms as primary and recommend that we adamantly resist the hidden agenda of what drives much of current postmodernist theory for fear of yet another label that has been thrust upon us.

The result is that some feminist scholars have tended towards a postcritical position that goes beyond critiquing the status quo. While critical theory was initially useful in pointing out the underlying struggles, some feminists (Luke and Gore 1992) have come to view critical theory as insufficient to differentially explain experiences for all women. In fact recent feminist literature has come to view critical and earlier feminist theory as adding to the perpetuation of its own master narratives. Hence the resistance to claims to view the world from a unified stance and to the acknowledgement of feminisms.

In virtually all of the feminist literature there is a focal interest in signification, in power-knowledge relationships, in the potential harm done by the master narratives both within and outside feminism, and in the way institutional structures are controlled. Recent feminist research in science education, however, has mainly focused on critiques of partriarchical assumptions and practices in efforts to document the politics and institutionalization of gender differences in educational settings (e.g., Guzzetti and Williams 1996; Roychoudhury, Tippins, and Nichols 1995). Middleton (1993) notes that explicit deconstruction of educational, theoretical master narratives has not been a visible activity in education. Luke notes that "the 'great' Western theoretical secular and non-secular master narratives which have been expressions of and the search for solutions to problems rooted in male experience" (1992, 29). The great need for the few feminist researchers in science education to align themselves with constructivism is an obvious illustration of this point about the alignment with master narratives. Having done so myself, I recognize the overwhelming influence to do so. For without an autobiographical analysis of my practice I would not be aware of such influences. Also, a feminisms position was particularly useful in understanding the influence of theorized feminism. The autobiographical process helped me to arrive at this position, to move outside the sacred stories of theorized feminism.

Recognizing the challenges of dealing with these dilemmas within the context of my practice, I will briefly outline summaries of my narrative inquiries below. The data used to prepare these summaries consisted of videotapes, interviews, research reports, presented papers, journal entries, lesson plans, action plans, formative and summative evaluations of courses, peer evaluations, teaching evaluations, and my own insights/musings that I have collected through the storying and restorying of my practice. To substantiate the credibility of my interpretation, I have engaged in prolonged data collection and analysis and persistent observation of my practice and have received extensive peer feedback at conference presentations.

I will now begin by responding to the first challenge that informed the analysis of my practice: the need to deal with the sacred stories of university practice, stories which contribute to maintaining the established power-knowledge relationships. Earlier I (Parsons 1999) described my interpretation of those years (1991–97) as a play with three acts where during each act I was played by a somewhat different feminist character. I would like to point out to the reader that the development is not as linear as it appears. The difference is that in this most recent story telling I will insert postcard entries at the end of the discussion of each act. These postcards represent my present reflections on my practice at that time.

Challenge # 1: Sacred Stories of University Practice in Three Acts

Act 1: A Historical Feminist Story (the Pre-SJSU Years)

When I arrived at San José State University (SJSU) in the fall of 1991 I began a feminist autobiographical analysis of my practice as a science educator (Parsons 1992). At that time it was important for me to understand where I was in my thinking. I started with an analysis of my published works, covering a period from 1987 to 1991. All the analyses presented in published papers were based on an extensive database. The following is a summary that traces my evolution through first-, second-, and third-generations of feminism as described by Kristeva (1982).

First-Generation Feminism. At the Fourth Girls and Science and Technology (GASAT) Conference, I presented a paper called "Females and Physical Science: Is Tinkering an Issue" (Parsons 1987). This was my first attempt to explore feminist issues in science education. The feminist position I took can be described as first-generation. While my reason for exploring tinkering was primarily to understand how to improve instruction for females in the physical sciences, I directed all my energy to understanding an activity in which males have typically excelled. The main argument I presented for doing so focused on having females acquire actional knowledge in the physical sciences, such that they could equal the males in achievement and participation. Such an argument did not look at broader issues, such as the nature of science, but indirectly laid the fault with the females for not having acquired such knowledge. Such a feminist position is representative of first-generation feminist thinking and was the feminist position I held at the beginning of my doctoral studies.

Second-Generation Feminism. In 1989 I co-authored a paper entitled "Sex-Related Differences in Science Achievement: A Possible Testing Artifact." My contribution to the work grew out of a review of science education literature where I noted that most of the discussion of the gender issue had centered around achievement on large-scale achievement tests (Bateson and Parsons 1989). Given that achievement had received so much attention, when I was presented with the opportunity to examine such data I was happy to reanalyze it from a feminist perspective. My major role in writing the paper was that of providing a theoretical framework for data analysis.

My thinking at the time suggested that I had moved beyond a liberal feminist stance to question the androcentric knowledge base of standardized achievement tests, a radical feminist position. While I introduced feminist theory to question the knowledge base of standardized tests, I still resorted to putting forward the differential experience of males and females as the primary issue, rather than challenging what is valued as scientific knowledge. I relied on the master narratives presented in the science education literature on the nature of scientific knowledge. Beyond raising the issue I did nothing to advance the discussion beyond a second-generation feminist perspective. My dominant focus was on using the argument of differential experiences to explain gender differences rather than challenging what knowledge has value in science. That is, I relied on studies which suggested that differential experiences of females and males are important in explaining gender differences in achievement in science. While this was a good rationale for changing test construction, it was also a good argument for rethinking school science, the case for which I made only a weak argument.

Third-Generation Feminism. Another illustration of second-generation feminist thinking is reflected in my paper "Female Participation in the

Physical Sciences: The Apprenticeship Factors" (Parsons 1991a). This paper was written before the Sixth International Gender and Science and Technology (GASAT) Conference and appeared in the conference contributions. At the conference, however, I felt the need to critique my earlier contribution. I critiqued my research for focusing on activities in physical sciences where males typically excel and, hence, I was blaming the victim. Why had I not focused on an area of science where females typically excel and used that as a basis for analysis? Another question that I raised was, Why not examine what is valued as scientific knowledge itself? While I was sensitive to the need to improve science instruction for females, I still laid the blame with the victim. This was despite the fact that feminist literature enabled me to describe such activities as disconnected knowing (Belenky et al. 1986). This realization signaled that I was ready to entertain third-generation feminist thinking when I arrived at SJSU.

During this pre-SJSU period it is important to note that constructivism also heavily influenced my work as a science educator. Two papers which illustrate my thinking at that time are "Making Sense of Constructivism in Preservice: A Case Study" (Parsons 1990) and "Preservice Secondary Science Teachers Making Sense of Constructivism" (Parsons 1991b). While my primary interest was feminism, the master narratives in science education (my discipline) did little to advance my thinking. During this time period I felt quite comfortable using constructivism as a master narrative to explore feminist issues.

Postcard #1: In Search of Feminist Theory
My experiences with university practice during act 1 were those that prepared me for the academy. I was introduced to the importance of theory in the academy. The two theoretical camps that influenced me were constructivism and feminism. I came to the university because my interest was in advancing females in science. I soon learned that the practical knowledge that I brought with me would only be good once I acquired indepth theoretical knowledge. The theoretical knowledge that I desired most was feminist theoretical knowledge. Feminist theoretical knowledge was also hard to get within science education; I was therefore attracted more to sociological literature. At that time Kristeva's (1982) work appeared to give me an interpretive framework. Later I found myself disagreeing with the use of various typologies and taxonomies to describe feminism. Middleton (1993) has similarly described her disenchantment with the use of first-, second- and third-wave feminist typology.

Act 2: An Emerging Postmodern Story—a Focus on Understanding the Needs of Teachers (1991-1993)

By restorying (Connelly and Clandinin 1988, 1991, 1993) my practice prior to 1991 I was made aware that in my first years at SJSU I was working towards moving my practice to a third-generation feminist perspective. I was also made aware that I was quite embedded in assuming a sociocultural perspective within constructivism (Parsons 1991a, 1991b). This was not surprising given that I had been a serious student of constructivism, having graduated from the University of British Columbia, with post-doctoral experience at Florida State University. Both sites were well known for their work in constructivism.

Earlier experiences had set the stage for me to bring radical and postmodern feminist perspectives to my practice as a science educator. Those prior experiences made me aware that being a feminist in science education is marginal. I was also aware that being a science educator in a predominately female college of education was equally isolating. SJSU, like other institutions, despite the fact that it was a college with an overwhelming number of female faculty, was still influenced by the same master narratives in education experienced elsewhere (Middleton 1993).

Both my prior experiences and my new institutional home would come to greatly influence my two interrelated research agendas—autobiography (a self-study of my practice) and action research (a collaborative inquiry into my work with preservice/inservice teachers).[1] It is also important to note that other than feminism and constructivism my work was influenced by critical theory. It was during this time that I started to explore action research from a critical perspective (Carr and Kemmis 1986). In my first autobiographical paper (Parsons 1992) I critically examined my work by beginning with a critique of constructivism. When I started autobiographical research I had no indication that this would become the primary focus of my research.

Postcard #2: In Search of a Feminist Epistemology of Practice

During my early years as a university science educator I undertook collaborative teaching as research work with my elementary methods preservice teachers. This was done through the use of constructivist methodology, which appeared to make a natural link with postmodern feminism. My students engaged in constructivist learning projects as part of the course work (i.e., interviewing students about science concepts, analyzing their teaching of a constructivist lesson). While the students completed the course assignments successfully, I sensed that many found science stressful. Deep down, however, I felt that the use of constructivist teaching methodology in science methods possibly even added to their stress. The fact that my students could intellectually understand and accept constructivism as a postmodern position did not mean that they felt comfortable with it. Many of my students, especially young women without teaching experience, felt alienated from not only educational theory but feminist theory. In fact theory did not appeal to the majority of my students. Even if theory now appeared to offer them an entry to science, science was something with which they felt very alienated. My students did not see themselves in the curriculum. I could not see my feminist self in the science education literature or in my teaching. Feminist knowledge to me at that time came from elsewhere. Postmodern feminism appeared to help me make sense of their stress and from where it was coming. My autobiography would hopefully help me to get to that elsewhere.

Act 3: A Story of Feminisms—a Science Education Learning Community Story (1993–1997)

By 1993 I became involved in an examination of the challenges in developing a collaborative relationship among student teachers, classroom teachers, and university faculty. A pluralistic, theoretical perspective informed by various feminisms influenced my work. I will now summarize some of the highlights of my action research work (1993–1997) with preservice/inservice teachers.

Cycle 1 (Spring '93–Fall '94): My Reflections on Initiating a Collaborative Relationship

Summer 1993 I presented at GASAT two papers which indicated a postmodernist influence on my practice (Parsons 1993; Parsons, Delauter, and De La Torre 1993). From that point postmodernism became a strong influence in my work. I assumed a postmodern position by focusing on voice and empowerment issues in my action research work with preservice/inservice teachers. Given that the focus in the first cycle (Parsons 1994) was on teacher preparation, it did little to create anything close to an emancipatory action research climate for the inservice teachers (Carr and Kemmis 1986). However, I recognized that we could not achieve a new model for science teacher preparation grounded in practice without teachers becoming partners in the process. It was evident that extended time was needed to establish an emancipatory action research climate at the classroom level. Also, it required a shift from teachers aiding in the establishment of the science-emphasis (technical-practical) program to teachers shaping how such a program might develop (emancipatory). In addition, the activity needed to move from being *theory driven* to *theory generating* in its orientation.

Cycle 2 (Fall '94–Spring '95): My Reflections on Establishing a Collaborative Relationship and a Community of Co-Learners

During the second research cycle we saw the beginnings of both a collaborative relationship and a community of co-learners (Parsons and Reynolds 1995). With the establishment of particular elementary school sites where science was to become an emphasis, the attempt to move towards greater teacher empowerment had been initiated. We had learned that to achieve in-depth collaboration we might need to limit our work primarily to certain school districts, and even to certain schools within those districts.

Our efforts had previously been focused on the preservice component and the establishment of science-emphasis school sites. However, we now needed to focus on defining inservice teachers' needs. A collaborative framework had begun, but it was delicate. Efforts needed to focus on the collaborative development of a future agenda based on common needs. Both the university and the schools shared a common link in preservice education, but we needed to redefine this common ground. Inservice education also had a shared focus at both the university and school levels. While we did not need to reach a consensus on these agendas it was important to establish some common ground. Becoming co-learners in redefining teacher preparation was viewed as part of professional lifelong learning for all participants.

The following selected passage from Parsons, Matson, and Rohan summarizes my thinking at the time: "The question is what strategies can we employ in the science education of adult women in preservice programs? We need to examine the profiles of elementary teachers who become interested in science because their voices need to be heard. Postmodern feminists (Nicholson 1990) recognize the need to listen to multiple voices if we are truly interested in bringing in the outsider (female elementary teacher) to science. Brickhouse (1994) uses the word *outsider* to describe underrepresented groups in science. In 1992 I undertook collaborative research with two preservice elementary teachers, Cathy and Blanca, in an attempt to share their stories (voices) with others who were interested in the empowerment of elementary teachers in science (Parsons, DeLauter, and De La Torre 1993). Fortunately, the undertaking of such collaborative research has allowed me for the first time to seriously listen to voices other than those of established science educators" (1995, 15). This and earlier works clearly illustrate my partial entrapment within the master narratives of science education. While I attempted to address the issues of voice and empowerment, postmodernism was becoming an accepted narrative within science education. At the time of the above writing I did not take seriously the caution within feminism of an alignment with mainstream theoretical camps such as postmodernism. It was later through the rereading of Lather's (1991) work, and works like Luke and Gore (1992), that I began to look at my work from a post-critical stance. It was through an awareness of feminisms that I started to bring a more pluralistic interpretation to my work.

A post-critical feminist stance meant that as a science educator, I must continue to struggle. As a feminist educator, I attempted on a daily basis to create pedagogical situations which empower all students, demystify canonical knowledge, and clarify how relations of dominate and subordinate subjects are marked by gender, ethnicity, race, class, sexuality, and many other markers of difference. There was, however, a serious caution to my good intentions. Luke and Gore specifically take "issue with the 'technology of control, and the silent regulation', indicated by such signifiers as 'power', 'voice', 'democratic freedoms', and the 'class, race, gender' triplet." (1992, 4)

In undertaking my work I struggled with the fact that from a feminist perspective the issue of empowerment is problematic. Gore highlights the problematic presuppositions and unreflective use of empowerment literature. Some of the problematic presuppositions which she identifies are as follows: (1) The fact that empowerment is a process which requires an agent (someone or something to empower) and even the notion self-empowerment presumes the agent self. (2) The notion of power as property—something the teacher (science educator) has and can give to the students (preservice/inservice teachers)—is often associated with zero-sum understanding of power. It is based on the belief that if teachers (science educators) give some of the power to students (preservice/inservice teachers), then they must give up some of their power. (3) The perpetuation of a dichotomy between empowerment and oppression also stems from a shift in conceptions of power as repression to power as productive, such that empowerment is linked with a productive conception of power and oppression is linked with a repressive conception of power (1992, 56–61).

In terms of the unreflective use of empowerment Gore's major concern stems from the issue of who serves as the agent of empowerment. Given that the agent is usually the teacher (or in this case a science educator), and that the subject (or object) of empowerment is others (teachers), a distinction is implied between them and us. Therefore there is danger apparent in the work of academics, such as myself whose discourse is purportedly empowering for the teachers. In focusing on others there is also a danger of forgetting to examine one's own implication in the conditions one seeks to affect (Gore 1992, 61). If empowerment is constructed as the exercise of power in an attempt to help others to exercise power (rather than the giving of power), I must confront the unforeseeable and contradictory effects of the exercise of power and must be humble and reflexive in my claims. Therefore I need to focus on examining the success of this academic at achieving the goal of a science education learning community rather than on a critique of others.

Cycle 3 (Fall '95-Spring '96): My Reflections on the Search for Common Ground

By the end of cycle 3 I recognized that while we had made more significant gains our long term goal, to establish a true collaborative relationship with schools, would take a long time (Parsons 1996). The building of such a relationship would require full support at participating school site levels and at the university level. I knew that if I wanted to learn more about how to make this process work better I would have to go with the sites that were the most promising, or what I would call hopeful sites.

I also felt a need to deal with the sacred stories of university practice. Connelly and Clandinin (1992) note that university faculty are central characters in the sacred story of expert knowledge relationships with schools. Rarely do university researchers (such as myself) see that our work is embedded in study and is also under study when we engage in work with schools. Science education research had focused heavily on inservice teachers improving their practice (Tobin et al. 1991), and in some cases on preservice teachers (Abel and Roth 1994; Martens and Crosier 1994). While such research has been respectful of teachers as professionals and sensitive to their needs, it has typically been theory driven, with the university science educator being in the position of power. This power imbalance contributed to the lack of reflection by the university faculty on their own practice. Kagan and Tippins (1991) suggested that university faculty work in a supportive rather than dominant role. If we want to create such learning communities, then both universities and schools needed to collaboratively restructure the teaching and learning process. We need to restructure the process, and author such as Greenleaf (1995) point to the need for the nurturing of new learning communities.

I felt there was a focal interest in signification, in examining the power-knowledge relationships, in the harm done by the master narratives, and in the way institutional structures are controlled. Therefore, in my work with teachers there was a need for sharing of power, voice, and ownership as we began to establish new ways of working in schools. My search for common ground is also informed by multiple voices (Ellsworth, 1992). Gore notes that rather than make pronouncements about what we can do, we need to ask, "What can we do for you?" (1992, 62). I would also add, What can we do together? The ongoing action research work that I undertook was but one small step in that direction. This search for common ground, however, all happened at a time when there were external pressures for collaboration between school districts and universities. My work with teachers suggested that forced (asked for) collaboration would not work.

Cycle 4 (Fall'96–Spring '97): My Reflections on the Continued Search for Routes to a Common Ground— Dealing with the Sacred Stories

While a preliminary analysis of the accomplishments looked great on paper, I knew the real struggle for change needed to begin (Parsons 1997). I really sensed that teacher education—as we had known it—was starting to be reconfigured. Alternative routes to teacher preparation were starting to appear. While I felt excited about this potential, I knew the huge challenges ahead. Many of my colleagues would resist this process because of a need to preserve the sacred rites within the university. Also, I knew from my own work that a tremendous effort would be needed to develop relationships where little history existed. The fact that we were encouraged to move ahead with the creation of university-school partnerships was also becoming a very political act. This need to create partnerships would automatically cast suspicion on the process. It was clear to me that teacher education needed to change, but how this change happened was critical. I sensed that educational history was being written. Some of us sensed that we were on the verge, where there was no new model yet.

I also felt that those of us who were interested in true reform of teacher education could learn from what had already happened. Stoddard (1993), in a discussion of some unsuccessful collaboration activities within university-school partnerships in professional development schools, noted that we need to find ways to work together to find a practical theory of pedagogy. However, to achieve this aim we need to overcome the obstacles of dissonance not only between university and school culture, but also between groups of faculty members within colleges of education. While Fedock, Zambo, and Cobern (1996) had focused on the need to bring scientists into schools to understand the work of science educators, we had forgotten that some of our colleagues in the colleges of education needed the same experience. My work suggested that if colleges of education did not start to value work in schools then the gap between the cultures would never be bridged. While it is important for science educators to reach out to the scientific community, it may be more important that we also overcome the political hurdles within our own community. It was important to work within to change the institutional structure before it was changed from the outside. From what I had seen, kindergarten through twelfth-grade (K-12) schools appeared to be responding to change faster than universities. This imbalance needed to be changed. Far too often the university, being the power-knowledge broker, asked the schools to change without undergoing change itself. A true university-school partnership would involve both partners fully participating in change over time. The employment of an action research agenda with an autobiographical connection carried with it a responsibility for transformative action, at both the school and university levels. Feminist theory had much to offer in bridging this gap.

Hollingsworth (1994) in her longitudinal action research with beginning teachers introduced the concept of "collaborative conversations" as a way of making sense of and improving practice. Similarly, in my work with teachers it is the longitudinal cycle of collaborative activities that has been continually examined to provide new directions for the improvement of my practice as a science educator. The focus of my analysis however has been on an examination of the sacred stories associated with my role in the process. I will acknowledge that I am only sharing one side of the story. There is another aspect of the story that needs to be shared, and that is the teachers' side of the story (Toolin 1995; Davis 1996). We need to get to the point where teachers are able to share and have a voice in the interpretation. We need to start engaging each other in sustained and connected conversation.

Postcard #3: Discovering the Limits of Theorized Feminism

As I became more and more involved in working with preservice/inservice teachers outside the traditional university context, I found feminist theoretical interpretations to be useful. Theorized feminism was particularly useful because there was limited educational research to draw from about school and university partnership work. This led me to the recognition that multiple feminist lenses could be applied in any one situation and would eventually lead to a feminist position. I also found that by engaging in a feminist autobiographical analysis I had discovered feminisms, and my evolving understanding of feminisms helped me to deal with the sacred stories of university practice. I wanted to understand my own situation as a feminist and a teacher. My questions were now coming from inside what I was studying. While the last statement may appear odd for someone engaged in an autobiographical analysis, I would argue that autobiographical research that is conceived within a theoretical context runs the risk of being controlled from the outside. It was through my recognition of the limitations of academic feminist theory that I gained the space I needed to work on the inside. My aim was that my students would be able to see that we were positioned inside what we were studying. My students have helped me over the years to formulate the analysis that I have described. Over the years they have continued to observe how my classroom continued to reproduce the same power relations described in the literature. Barton notes that challenging our structural conditions is so "difficult because they permeate everything we say, do, and experience" (1998, 33).

Next, I will explore my second challenge: dealing with the sacred stories of theorized feminism. In my exploration of this challenge I would like to borrow from Middleton who notes, " My feminism and educational theories inform my public and private life outside the university, and my feminist community activities influence my curricula" (1993, 103).

Challenge # 2: The Sacred Stories of Theorized Feminism

The concept of voice began appearing in my writings around 1993 (Parsons, Delauter, and De La Torre 1993). When I used voice in my earlier writings it was an acknowledgement of different voices from a theoretical perspective. Then later (1998 onwards), by exploring "angels in the house" I came to understand what it meant to write in one's own voice (Parsons et al. 1999).[2] The next section summarizes my thinking.

Angels and Other Voices

By starting with a classic action research agenda and at the same time beginning an autobiographical analysis of my own practice from multiple feminist perspectives, by 1997 I had come to a point where both research agendas had merged (Parsons 1998). I had also come to critique my own quest to help facilitate the empowerment/emancipation of others. Nicholson (1990) notes that differences in class, race, culture, and sexual orientation suggest that multiple feminist voices should be heard within the learning community and academe. In both cases we needed voices rather than a dominant feminist voice.

I was also made aware of the need to examine issues related to my own empowerment within a university context. By focusing on the conditions needed to improve the practice of others, I had come to understand what was needed to improve my own practice. Being a female, an immigrant, and being from working-class background did not give me ready access to the American university power structure where I worked. Middleton, however notes that while many of today's feminist educators were among those who often felt marginalized within academe, we often had sufficient cultural capital "to enable our marginalization to become the basis of an intellectual critique rather than of educational withdrawal or failure" (1993, 4-5). Therefore, within the context of academe I recognized that from the point of view of a feminist praxis I needed to kill the angel in academe (Hollingsworth, 1996).[3] To do this I needed to deal with the existing power-knowledge relationships within academe. I had come to realize that what is empowering for me might not be empowering for others. I had also come to realize that care needed to be taken not to add to the development of feminist power-knowledge relationships in science education.

Then Later, in spring 1999, when I read Brookes's (1992) Feminist Pedagogy: An Autobiographical Approach, I came to understand what it meant to write in one's own voice. Brookes writes, "In the early stages of learning to critique socially organized illusions, I found it difficult to trust my own words or perspectives. I was like a child learning to read and write anew. For this reason, I very consciously borrowed the words of others. I chose them because they acted as a kind of viewfinder for me. In other words, I borrowed language to see through my own experience, story, perspective" (1992, 5). Brookes's work illustrates how easy it is to hide behind theory and let theory become our voice. While theory gave me an appreciation and sensitivity for different voices, I believe it is by coming to understand the differences in my own voice that I will truly begin to appreciate the differences in others voices. I also see a difference between using theory and hiding behind theory. I would like to think that the recognition of feminisms has allowed me to use theory rather than being blinded by theory.

Venturing beyond the Autobiographical: Women's Voices in the Eastside Story

In spring 1999 I also came together with four other educators to explore issues involved in our work to develop a science education learning community that addresses not only science but literacy for first- and second-language learners (Parsons et al. 1999). We called our story the "Eastside Story" for two reasons: (1) most of our work takes place in East San José, which serves the largest percentage of underrepresented students in the San José area; and (2) we associate our work with challenging the existing power-knowledge structures within education.

The collective theoretical position taken in our work has been described as feminisms. There is no one position in feminist theory that informs all of our work. While some of us have been influenced by postmodern feminism, we find it inadequate to describe our collective epistemological position. We are aware that the central tenant of postmodernist theory is an anti-foundational epistemology which rejects foundational truths in disciplinary knowledge and the unitary rationalist subject as foundational to all knowledge (Luke and Gore 1992; Nicholson 1990). While collectively we can agree with the basic tenants of postmodern feminism, it is not inclusive of all our views. Our voices in the Eastside Story have therefore assumed pluralistic, theoretical perspectives, influenced by various feminisms.

We have drawn from works such as Nicholson (1990) to inform our collective position. Our differences suggest that multiple feminist voices need to be heard, rather than a dominant feminist voice. We recognize that while the postmodern position is not a perfect match for feminist theory it is, nevertheless, a workable set of identity papers that allows some feminists to communicate with a broader audience.

The result is that we prefer to critique the status quo and to deconstruct the master narratives. We identify with the fact that the feminist literature focuses interest in signification, in power-knowledge relationships, in the potential harm done by the master narratives both within and outside feminism, and in the way institutional structures are controlled. We also find ourselves in agreement with the feminist literature that has come to view earlier feminist theory as adding to the perpetuation of their own master narratives (Luke and Gore 1992). Such a position has been described as post-critical which is the resistance to claims to view the world from a unified feminist stance through the acknowledgement of feminisms. While this position may be viewed by some as a paralyzing act, it is an acknowledgment of the usefulness of multiple frames of reference. Collectively we represent these multiple frames. Given the diversity of the issues associated with our work, no one position will do.

Voices in the Eastside Story

We will now step into our role as educators and share our stories.[4] By sharing our stories of literacy through science, a feminisms theoretical perspective will be assumed. This allows each of us to communicate in our own voice about our work. Very abbreviated versions of those voices are summarized here: Rosalinda , a bilingual educator who entitled her work *Lonely Voice in Academe* said: "Importantly, I am the only Latina tenure track professor in the teacher education department. I was hired eight years ago, and no other Latino/a professors have been recruited as tenure-track faculty. There is one other Latino male in teacher education, although he has been more involved in the administrative part of the teaching program."

Norma, a bilingual teacher who entitled her work *Awakenings* said: "For a long time, living to read was my reality, for I once believed that knowing how to read was equated to having power. Now, I've come to understand that reading someone else's reality, is just not as real, nor as powerful as when you engage in writing to construct your own understandings about your reality in the process of making sense of it. That is power!" Carol, a middle school science teacher who entitled her work From Nerds to Now said: ". . . I want the female students in my class to know that science is not only for nerds." Merle, a high school science teacher who entitled her work Three Strikes and I Am Out / Female, Black and Third World said: ". . . I think it took the staff so long to value my opinion because I was female, black and from the third world."

It has been through the Eastside Story that we have discovered the importance of multiple voices to address the important issue of literacy for English-language learners (for both first- and second-language learners) in science. It has been through our work that we have come to view literacy as an equity issue for science educators. We have also come to view literacy as essential for the success for today's under-served students in urban settings.[5]

Addressing the Sacred Stories in My Practice

Through my ongoing inquiry, I have come to better understand some aspects of the various influences on my practice. I have also come to value the use of narrative inquiry as an autobiographical technique. It has been through an ongoing feminist examination of my practice that I have become aware of the role of master narratives within science education and feminism.

My autobiography suggests that, with regards to bringing a pluralistic perspective influenced by various feminisms to my practice, I now need to challenge the dominant narratives within both science education and theorized feminism. To do this, I need to move beyond the level of a discussion of feminist issues reported in most science education and feminism.

Another outcome of my narrative inquiry is to question the future direction of research into the practice of all science educators. Given that my research has been influenced by feminist epistemology, I certainly sense the need to examine the nature of science education with a focus on how can we develop a new focus in the reform of science education of all teachers (preservice, inservice, and university educator). In attempting to accomplish such a task my autobiography has illustrated two of my ongoing challenges. As a feminist science educator I have no choice but to explore the issues of working within, and against, the normalizing borders of science and science education and working towards a feminist image of science education (Lather 1994). To this dilemma I would also add the complexities of dealing with the dual worlds of the school and university, and with the sacred stories within my own institution. Through my ongoing autobiographical analysis I have discovered that my practice as a science educator is a multifaceted interface which involves the following:

- Exploring ways of dealing with feminisms within the context of science education;
- Finding ways to a common ground between the world of K-12 schools and the university;
- Developing a feminist praxis within the context of sacred stories of university practice; and
- Using multiple voices to inform my practice.

In all four instances, I feel an overwhelming pressure to conform to the master narratives; in all four instances the very essence of my work is to change these narratives. In attempting to work within my discipline and attend to the feminist literature, I have often fallen into the trap of not critiquing the status quo within my own discipline, or within feminism. Within my own institution I have also fallen into the role of listening to the angel rather than challenging the status quo (Hollingsworth 1996). Through my inquiry I have found ways to reshape my practice and not be overcome by the pressures of the status quo. I will continue to do so in the future not only by bringing multiple feminist perspectives to an examination of my practice but by utilizing multiple voices to inform my analysis.

My autobiography illustrates the need for university science educators to continually reflect on their practice. Connelly and Clandinin (1992) note that we also need to critically reflect on the university's role in the process. I believe that we need to do both. Science educators generally work within the paradigm of "normal science education" to play on Kuhn's (1970) words (Parsons, Matson, and Rohan 1995). We need to step outside our paradigm if we are asking teachers to do so. The employment of a feminist agenda carries with it a responsibility for transformative action at both the school and university levels.

My ongoing autobiographical analysis has not only informed my work, but is transforming my practice as a science educator in ways that I could not have imagined. It has also continued to inform my position within feminism. An exploration of the various theoretical perspectives and my eventual questioning of theorized feminism have allowed me to bring multiple voices to challenge the master narratives within my practice as a science educator. Each time I critically analyze my practice I come to see my practice anew.

References

Abell, S. K., and M. Roth. 1994. Constructing science teaching in the elementary school: The socialization of a science enthusiast student teacher. *Journal of Research in Science Education* 31 (1): 77–90.

Barton, A. C. 1998. *Feminist science education*. New York: Teachers College Press.

Bateson, D., and S. Parsons-Chatman. 1989. Sex-related differences in science achievement: A possible testing artifact. *International Journal of Science Education* 11 (4): 375–85.

Belenky, M. J., B. M. Clinchy, N. R. Goldberger, and J. M. Tarule. 1986. *Women's ways of knowing: The development of self, voice, and mind*. New York: Basic Books.

Brickhouse, N. 1994. Bringing in the outsiders: Transforming the sciences. *Journal of Curriculum Studies* 26 (4): 401–16.

Brookes, A. L. 1992. *Feminist pedagogy: An autobiographical approach*. Halifax, Nova Scotia: Fernwood Publishing.

Bullough, R. V., Jr. and A. Gitlin. 1995. *Becoming a student of teaching: Methodologies for exploring self and school context*. New York: Garland Publishers.

Carr, W., and S. Kemmis. 1986. *Becoming critical: Education, knowledge and action research*. Philadelphia, PA: Falmer Press.

Connelly, M., and J. Clandinin. 1988. *Teachers as curriculum planners: Narratives of experiences*. New York: Teachers College Press.

———. 1991. Narrative inquiry: Stored experience. In, *Forms of curriculum inquiry*, ed. E. Short, 121–54. Albany: State University of New York Press.

———. 1992. The promise of collaborative research in the political context. Paper presented at Annual General Meeting of American Educational Research Association, April, San Francisco, CA.

———. 1993. Narrative inquiry. *Educational Researcher* 19 (5): 2–14.

Cortazzi, M. 1993. *Narrative analysis*. Bristol, PA: Falmer Press.

Cuban, L. 1992. Managing the dilemmas while building professional communities. *Educational Researcher* 21 (1): 4–11.

Davis, N. T. 1996. Looking into the mirror: Teachers' use of autobiography and action research to improve practice. *Research in Science Education* 26 (1): 23–32.

Ellsworth E. 1992. Why doesn't this feel empowering? Working through the repressive myths of critical pedagogy. In *Feminisms and critical pedagogy*, eds. C. Luke and J. Gore, 54–73. New York: Routledge.

Fedock, P., R. Zambo, and W. W. Cobern. 1996. The professional development of college science professors as science teacher educators. *Science Education* 80 (1): 5–19.

Giroux, H. 1992. *Border crossings: Cultural workers and the politics of education.* New York: Routledge.

Gore, J. 1992. What can we do for you! What can "we" do for "you"? Struggling over empowerment in critical and feminist pedagogy. In *Feminisms and critical pedagogy*, eds. C. Luke and J. Gore, 54–73. New York: Routledge.

Graham, R. J. 1991. *Reading and writing the self: Autobiography in education and curriculum.* New York, NY: Teachers College Press.

Greenleaf, C. L. 1995. You feel like you belong: Student perspectives on becoming a community of learners. Paper presented at the Annual General Meeting of the American Educational Research Association, April, San Francisco, CA.

Guzzetti, B. J., and W. O. Williams. 1996. Gender, text, and discussion: Examining intellectual safety in the science classroom. *Journal of Research in Science Teaching* 33 (1): 5–20.

Harding, S. 1986. *The science question in feminism.* Ithaca, NY: Cornell University Press.

———, ed. 1987. *Feminism and methodology.* Indiana: Indiana University Press.

Hildebrand, G. M. 1995. Re/Viewing gender and science education via multiple frames of reference. Paper presented at the Annual General Meeting of the National Association for Research in Science Teaching, San Francisco, CA.

Hollingsworth, S. 1994. *Teacher research and urban literacy.* New York: Teachers College Press.

———. 1996. Killing the angel in academe: Feminist praxis in action research. Paper presented at AERA Research on Women and Education, October, San Jose, CA.

Kagan, D., and D. Tippins. 1991. How teachers' classroom cases express their pedagogical beliefs. *Journal of Teacher Education* 42 (4): 281–91.

Keller, E. F. 1985. *Reflections on gender and science.* New Haven, CT: Yale University Press.

Kristeva, J. 1982. Women's time. In *Feminist theory: A critique of ideology,* eds. N. O. Keohane, M. Z. Rosaldo, and B. C. Gelpi, 31–54. Chicago: University of Chicago Press.

Kuhn, T. S. 1970. *The structure of scientific revolutions.* 2nd ed. Chicago: University of Chicago Press.

Lather, P. 1991. *Getting smart: A feminist research and pedagogy with/in the postmodern.* New York: Routledge.

———. 1992. Post-critical pedagogies: A feminist reading. In *Feminisms and critical pedagogy,* eds. C. Luke and J. Gore, 120–137. New York: Routledge.

———. 1993. Fertile obsession: Validity after poststructuralism. *Sociological Quarterly* 34 (4): 673–693.

———. 1994. An ache of wings: Women aids and angels. Paper presented at the Annual General Meeting of the Conference on Curriculum Theory and Classroom Practice, October, Banff, Alberta.

Luke, C. 1992. Feminist politics in radical pedagogy. In *Feminisms and critical pedagogy,* eds. C. Luke and J. Gore, 25–53. New York: Routledge.

Luke, C., and J. Gore, eds. 1992. *Feminisms and critical pedagogy.* New York: Routledge.

Martens, M. L., and S. Crosier. 1994. Sharon's story: The usefulness of conceptual change constructs in promoting student reflection. *Journal of Science Teacher Education* 5 (4): 139–45.

Middleton, S. 1993. *Educating feminists: Life histories and pedagogy.* New York: Teachers College Press.

Nicholson, L. J. ed. 1990. *Feminism/Postmodernism.* New York: Routledge, Chapman and Hall.

Noddings, N. 1984. *Caring: A feminine approach to ethics and moral education.* Berekely, CA: University of California Press.

Parsons, S. 1987. Females and physical science: Is tinkering an issue? In *Girls and Science and Technology: Contributions to the Fourth GASAT Conference*, eds. J. Z. Daniels and J. B. Kahle, Vol. 1, 97–104. Ann Arbor, MI: University of Michigan.

———. 1990. Making sense of constructivism in preservice: A case study. Paper presented at the annual meeting of the National Association of Research in Science Teaching, April, Atlanta, GA.

———. 1991a. Female participation in the physical sciences: The apprenticeship factors. In *Gender and Science and Technology: Contributions to the Sixth International GASAT Conference*, eds. L. .J. Rennie, L. H. Parker, and G. M. Hildebrand, Vol. 1, 286–94. Perth, Australia: Key Centre for Teaching and Research in School Science and Mathematics, Curtin University.

———. 1991b. Preservice secondary science teachers making sense of constructivism. *Research in Science Education* 21: 271–80.

———. 1992. Underrepresented groups in science: The diary of a critical constructivist. Paper presented at the National Association for Research in Science Teaching Annual Meeting, April. Cambridge, MA.

———. 1993. The history of feminist theory as a basis for review of the gender and science question. In *Contributions to the Seventh International Gender and Science and Technology Conference*, ed. S. Haggerty and A. Holmes. Ontario, Canada: Ontario's Women's Directorate.

———. 1994. California State University- San Jose State Science Teacher Development Project. Paper presented at the Annual General Meeting of AETS, January, El Paso, TX.

———. 1996. Establishing a science education learning community: Preservice/inservice teachers and teacher educators as co-learners.

Paper presented at the Annual General Meeting of the National Association of National Research in Science Teaching, April, St. Louis, MO.

———. 1997. The search for routes to a common ground for the building of a community of co-learners. Paper presented at the Annual General Meeting of National Association of National Research in Science Teaching, April, Chicago, IL.

———. 1998. *A science education learning community story.* Columbus, OH: ERIC Clearinghouse for Science, Mathematics, and Environmental Education (ERIC publication ED 418 868).

———. 1999. Feminisms and science education. *International Journal of Science Education.* 21(9): 989–1005.

Parsons, S., C. Delauter, and B. De La Torre. 1993. In a different voice: Empowering female elementary teachers in science. In *Contributions to the Seventh International GASAT Gender and Science and Technology Conference, University of Waterloo,* ed. S. Haggerty. Ontario, Canada: Ontario's Women's Directorate.

Parsons, S., and J. O. Matson. 1995. Through the looking glass: An autobiographical study by two science educators. Paper presented at the Annual General Meeting of the Association for the Education of Teachers in Science, January, Charleston, WV.

Parsons, S., J. O. Matson, and J. Rohan. 1995. The art of reflecting in a two-way mirror: A collaborative autobiographical study by three science educators. Paper presented at the Annual General Meeting of the National Association of National Research in Science Teaching, April, San Francisco, CA.

Parsons, S., R. Quintanar, N. Rodriguez, C. Hagen, M. Boxill. 1999. Women's voices in the Eastside story. Paper presented at a Post-Annual General Meeting of theNational Association of National Research in Science Teaching on Gender and Science Education, April, Boston, MA.

Parsons, S., and K. E. Reynolds. 1995. Establishing an action research agenda for preservice and inservice elementary teachers collaboration on self-empowerment in science. Paper presented at the Annual General Meeting of the National Association of National Research in Science Teaching, April, San Francisco, CA.

Pinar, W. 1988. "Whole, bright, deep with understanding": Issues in qualitative research and autobiographical method. *Contemporary curriculum discourses*, ed. W. Pinar, 134-53. Scottsdale, AZ: Gorsuch Scarisbrick.

Pinar, W., W. Reynolds, P. Slattery, and P. Taubman. 1995. *Understanding curriculum*. New York: Peter Lang.

Robeck, E., G. Hepburn, and J. Gaskell. 1995. Re/presenting a gender equity project: Pedagogical approaches and methodological concerns. Paper presented at the Annual Meeting of the National Association of Research in Science Teaching, April, San Francisco, CA.

Roberts, H., ed. 1981. *Doing feminist research*. London: Routledge and Kegan Paul.

Rodriquez, A. 1995. "I'm just performing": Playing roles, resistance, and managing the dilemmas of learning to teach science. Paper presented at the Annual General Meeting of the National Association of National Research in Science Teaching, April, San Francisco, CA.

Rohan, J. 1996. *Making sense of experience: An autobiographical study*. Masters thesis, San Jose State University, CA.

Roychoudhury, A., D. J. Tippins, and S. E. Nichols. 1995. Gender-inclusive science teaching: A feminist-constructivist approach. *Journal of Research in Science Teaching* 32 (9): 897-924.

Stoddard, T. 1993. The professional development school: Building bridges between cultures. *Educational Policy* 7 (1): 5-23.

St. Pierre, E.A. 1997. Nomadic inquiry in the smooth spaces of the field: A preface. *Qualitative Studies in Education* 10 (3): 365-83.

Tobin, K., N. Davis, K. Shaw, and E. Jakubowski. 1991. Enhancing science and mathematics teaching. *Journal of Science Teacher Education* 2 (4): 85-89.

Toolin, R. 1995. Teaching science in elementary school: A narrative of a preservice teacher's perspectives about science and science teaching. Paper presented at the Annual General Meeting of the National Association for Research in Science Teaching, April, San Francisco, CA.

Tippins D., S. E. Nichols, and K. Tobin. 1993. Reconstructing science teacher education. *Journal of Science Teacher Education* 4 (3): 65–72.

Notes

1 Davis (1996) has similarly used these research techniques in her work with teachers.

2 Hollingsworth (1996), borrowing from Virginia Wolf, talks about the need to drive out those inner voices (the angel within us) which keep us conforming to the status quo.

3 Hollingsworth (1996) proposes that just as Virginia Woolf had to kill the angel in the house, as academics we also need to kill the angel within us to deal with the power-knowledge relationships within academe.

4 Since I have already shared my story in depth I will share the other four voices in the Eastside Story.

5 We prefer to use "under-served" instead of "students in low-performing schools," which is a classification now being used in California to describe low performance by students on standardized tests.

Chapter 12

Sociocultural Constructivism, Courage, and the Researcher's Gaze: Redefining Our Roles as Cultural Warriors for Social Change

Alberto J. Rodriguez

Laying Down Invisible Boundaries

In this chapter, I propose that we urgently need to reexamine our roles as researchers and teacher educators in order to effectively address the reproduction of social and educational inequalities in our schools. As a point of departure, I suggest we engage in deconstructing the notion of educators as cultural workers.

All those who teach and conduct research about teaching and learning are indeed cultural workers. This is due to the nature of our work, whether it is quantitative or qualitative research and whether or not we choose to acknowledge it. Our work in education affects the cultures of teaching and learning by either reproducing the status quo or by adding effective ways to improve it. The continuation of pervasive gaps in student achievement in science and mathematics in national tests for the last twenty years (Rodriguez 1998a; NSF 1996) and more recently in international tests (Schmidt 1998); the consistent gender differences in participation in science, engineering, mathematics, and technology-related fields (Rodriguez 1998a; NSF 1996); the consistently high dropout rate of Latinos/as for the past twenty-six years (Secada, et al. 1998); all of these clearly indicate that being cultural workers is not enough. We need to become *cultural warriors* for social change. The analogy of teacher educator/researcher as cultural warrior more directly acknowledges the deep impact our work needs to have on education—and more clearly points to the need to redefine it. This is, of course, no easy task, and it requires that we find the political will and courage to manage the risks associated with not only talking about social change but also being an integral part of its process.

This leads to the main question I wish to address in this essay, How do we help improve the preparation of our teachers and the achievement and participation of all students in science? My answer is, With courage. Courage has been the missing element in education reform in general, and in science education reform in particular. In the following sections, arguments are presented to illustrate that in the last twenty-five years of research in science education we have accumulated a great of deal knowledge about teaching and learning. This knowledge has primarily served *us* as a research community, but it has had little impact on how teachers teach, on how students learn, and on the existing social inequalities that continue to make teachers' work so impossible. Therefore, by redefining our roles as cultural warriors, we will find the courage to manage the risks and the resistance to change we will most certainly encounter once the ruling hegemony is challenged.

One (Auto)Biographical Note

When I got hired by one of the top ten universities in the United States upon completion of my doctorate in science education, I thought I would finally have a supportive space in which I could continue learning how to teach for diversity and understanding[1]. After all, I was thrilled about the opportunity to work with so many scholars whose work has encouraged and guided my own as a graduate student. I thought to myself, "Finally, I can spend more time *doing* the work rather than spending so much energy arguing with my peers and professors about why social justice work in science education is important." Unfortunately, it was not until my first day of classes that I realized that so little had changed since I made the decision to leave my teaching job and go to graduate school. Once again, there I was, the only person of color (and the only Latino) in the class. There it was again, the look. That powerful gaze privileged individuals have to let you know that they do not approve of you, that they can place you, and that they resent you being there. Without words, the "look" sends shivers down your spine because you know that you have just been placed at the bottom of the hill. You know that from that moment on it is going to be a lot of work to climb up—because that is the only way you can go when you are *placed* at the bottom.

That first day of classes was probably the most difficult day of teaching I have ever encountered as a new professor. The condescending silence of the students' body language dominated the air, exposing in my voice the desire to just run out of that room. As I continued going over the course syllabus, I reminded myself why I was there and why—more than ever—it was important for me—the only person of color in that room—to do my job as a science educator.

As the course went on, I began to make alliances with most of the students. Through careful reflection and honest dialogues, I began to understand why some students reacted to my presence in the classroom with such apprehension. It became apparent that for many of them I was the first person of color they have ever encountered in a teacher-student relationship—or in any relationship for that matter. This was the first time I realized that some of their hostility was rooted in their unspoken fear. There I was, somebody pushing for equity issues in science education, sharing my own experiences with oppression, but it never occurred to me that the power I had as a professor, combined with my cultural experiences, worried some of them. Some were worried because I was perceived to have a chip on my shoulder; hence, they thought they were going to be punished as a result of their privileged locations. Others strongly felt that they just came to my class to learn how to teach science, and that their future students' socioeconomic and cultural backgrounds were none of their concern. In the course evaluations, I often had several students lament about how multicultural education was being "pushed down their throats." This was disturbing to me. Why were some students complaining about the focus of our teacher education program? Didn't they know that world-famous scholars work in our department and that power issues, multicultural education, curriculum reform, equity, and so on are the topics about which they have written and continue to write extensively? I began to wonder why I was the only professor of color these students had encountered in their teacher preparation. I discovered that some of the few professors of color in our Department no longer worked with undergraduate students—and, similar to their Anglo (mostly male) colleagues, only taught graduate courses. In any case, I sought the advice of some of my colleagues. Surely, they had taught a methods class at one time or another in the past. What suggestions would they offer to a fellow professor to manage the preservice teachers' resistance to learning to teach for diversity? The common advise I got was simple—and disappointing: "Do not rock the boat." "Give the students what they want until you get tenure." "Once you get tenure, you can do your own thing in the classroom." This advice transported me back to my first teaching job in public school, where keeping my job meant keeping quiet and not challenging students to think about their potential to be agents of change. This was old advice that I did not follow then, and I was certainly not going to follow it now.

It became obvious that what I needed to do was to learn how to be a better teacher in a hostile and contradictory environment, and to use these experiences as a vehicle to make my research more socially relevant. I needed to continue being a warrior for social change—not just write and tell other people to be warriors, but through my own struggles in practice learn to be a more effective multicultural teacher educator. The alternatives were not real options. I would not compromise my principles, and just like when I was a public school teacher, I would not quit and follow someone else's agenda. I would go somewhere else when I was ready, and when I was satisfied with disrupting the contradictions in our multicultural teacher education program. A program that intellectually urged our future teachers to take all sorts of risks to effect change in their schools, yet it discouraged beginning professors rocking the boat in their own classrooms. We had a teacher education program that ironically was encouraging me to avoid the kind of conceptual conflicts that our own students were bound to encounter wherever they attempted to conduct the same kind of transformative work that some of us were preaching.

Through extensive student course evaluations and honest dialogue with students and colleagues from other universities, I became aware that one way to manage preservice students' resistance to teaching for diversity and understanding was implementing *strategies for counter-resistance* (Rodriguez 1998b). These strategies were not meant to antagonize—although that was sometimes unavoidable—but they were meant to create powerful moments of conceptual and cultural dissonance where students and instructor could see and appreciate each other's location in time. These moments created safe spaces for me and my students to critique our taken-for-granted assumptions and open up to new possibilities for learning, for teaching, and for working across multiple differences in cultural, socioeconomic, and historical locations.

As I continued exploring ways to become an effective multicultural science teacher educator in an often-hostile environment, I also looked for guidance in the relevant literature. What I found was frustration with the rhetoric of critical pedagogy and multicultural education because it did not provide me with specific suggestions on how to implement the multiple and demanding changes being proposed. On the other hand, teacher education research based on individual constructivism— the predominant theory of learning in the last three decades—provided various models and suggestions for enhancing teacher preparation and student learning, but ignored students' cultural, socioeconomic, historical, and institutional contexts as important factors influencing the construction of new knowledge. Therefore, in a recently published article in the *Journal of Research in Science Teaching* (Rodriguez 1998), I proposed sociotransformative constructivism (STC) as an orientation to teaching and learning that links multicultural education tenets with social constructivism. In that paper, I also explained the results of a year-long study with preservice teachers. The study focused on the effectiveness of various strategies of counterresistance for assisting students in their learning to teach for diversity and understanding. A more detailed example of STC and how I use it in my science methods classes will be explained below, but what we need now is to reconsider first the reasons why we should all become cultural warriors for social change.

The Courage to Gaze Inward

I have never let my schooling interfere with my education.
—Mark Twain

What was Twain's humorous cynicism then has now become a serious social issue today. Many students—especially Latino children—continue to dropout (or are pushed out) of school at an alarming rate. We have seen improvement in the drop out of rates of U.S. African and Anglo students, but the dropout rate of Latinos/as has virtually stayed the same for the last twenty-six years. In 1972, about 21 percent of U.S. Africans between the ages of sixteen and twenty-four years old had dropped out of school. The rate for the same ethnic group was 12.6 in 1994—a drop of almost nine percentage points in twenty-two years. In contrast, the drop out rate of U.S. Anglo students for the same age group changed from 12.5 percent in 1972 to 7.7 percent in 1994. On the other extreme, the drop out rate of sixteen and twenty-four years old Latino students has consistently remained between 30 and 35 percent for over two decades (Secada et. al. 1998; McMillen 1994)! In the meantime, demographic projections indicate that Latinos/as will be the second largest ethnic group in the U.S. by the year 2010 (Day 1993; Secada et. al. 1998). We are heading toward a social and economic crisis if we do not address the current trend of low achievement and participation of the fastest growing ethnic group in the country. Jeannie Oakes has showed us how social inequalities in schools tend to have a multiplicative effect throughout the students' schooling (Oakes and Lipton 1998; Oakes, Ormseth and Campbell 1990). We also know that poor parental education is a major factor on how children perform and whether they stay in school (Peng 1995; Romo and Falbo 1996; Secada, et al. 1998). Thus, it is no wonder that the cycle of poverty observed for those who drop out of schools is more pronounced in the lives of Latinos/as and their children. According, to the National Center for Education Statistics, "over half of the dropouts not pursuing any further education in 1994 had at least one child (as either a cause or a consequence of their dropping out)" (McMillen 1994, 51). In 1991, Texas, for example, held the unfortunate title of having the highest percentage rate of births to girls fourteen years old and younger in the nation (Romo and Fabo 1996). Latinas fifteen years old and younger constituted 50 percent of the teen mothers in Texas in the same year.

What are the causes for these trends? While some would prefer to blame the victim for being caught in a spiraling cycle of poverty, a substantial body of research indicates that factors such tracking, grade retention, standardized testing, lower teacher expectations, peer pressure, family environment, low socioeconomic status, and parents' low educational background among others contribute to the observed gaps in student achievement and participation in science (Rodriguez 1998a; NSF 1996). It is ironic indeed that our students—and particularly traditionally underrepresented students in science[2]—are in such a predicament today. It is ironic because, during the same last two decades, we have augmented our understanding of students' alternate conceptions on a variety of science concepts (Fensham, Gunstone and White 1994; Gabel 1994). We also have a better understanding of pre- and inservice teachers' beliefs about teaching and learning (Gabel 1994; Carter 1990; Zeichner 1990; Kagan 1992). Most of these studies have also been informed by individual constructivism—a theory that states that "children's learning is a process of personal, individual, and intellectual construction arising from their activity in the world" (Matthews 1994, 138). As a matter of fact, Pfundt and Duit (1991) compiled a bibliography of more than 1,100 studies informed by individual constructivism as the main theoretical framework.[3] To continue the irony, even though research based on individual constructivism has been very fruitful, the technorational approach of this framework appears to have mainly benefited the science education research community. It has had little effect on teachers' and students' lives in the classroom .

According to the National Survey of Science and Mathematics Education, a typical elementary or secondary science class spent:

- Almost 40 percent of its time in lecture and discussion involving the entire class.
- About 20 percent of its time working as individuals reading the textbook or completing worksheets.
- About 25 percent of its time working with hands-on materials.
- The remaining of the time on daily routines and non-laboratory small-groups. (NSF, 1996 60).

Furthermore, less than half of all high school teachers assign long-term projects, and most—who have computers—are not using them for instruction on a regular basis (NSF 1996). The dissonance between what individual constructivist research suggests and how teachers continue to teach—again—is ironic. More specifically, while two decades of research based on individual constructivism has increased our knowledge of students' prior conceptions on topics such as heat, light, and various biological processes, thousands of children do not have heat and light and have little or no food every night. For instance, according the 1994 National Commission on Children Report (Stallings 1995), every night at least 100,000 children are homeless in the U.S.; nearly 13 million children live in poverty—over 2 million more than a decade ago; and dropouts are 3.5 times more likely to be arrested than students who complete school.

Given the continuing reproduction of social inequalities in schools; given the pervasive gaps in student achievement and participation in science; and given the high drop out rates of Latino/a students, we must have the courage as a research community to gaze inward and ask ourselves:

- What impact has our research really had on how teachers teach and on how students learn for the last two decades?
- Whose interests are being served by our work as science teacher educators and researchers?
- How can we make better connections between our work and teachers' school practice, and students' lives?

We need to reflect on what our research has contributed to address these kinds of questions. If we have the courage to expand our conceptualization of what accounts as useful research, we will find, for example, that feminist post-structuralist frameworks are more focused on social transformation (Reinharz 1992). For instance, Patti Lather (1991) building on the work of others has suggested that we consider *catalytic validity* as an additional construct to assess the impact of our work with the Other. She explains that "catalytic validity represents the degree to which the research process re-orients, focuses and energizes participants toward knowing reality in order to transform it, a process Freire terms conscientization" (1991, 68).

Feminist poststructuralist frameworks can also help us avoid the conceptual traps embedded in our positions of privilege by seeking to implicate ourselves—reflexively—in the process of change. In this way, we can also avoid leaving the Other—the researched—to be the one doing all the transformative work while we retreat to the safety of our offices. Mimi Orner (1992) makes this point clear when she states that "Feminist poststructuralist theories offer powerful tools for analyzing the mechanisms of power locally and the possibilities for change. In education, the call for voice has most often been directed at students. Where are the multiple contradictory voices of teachers, writers, researchers, and administrators?" (p. 88).

Another alternate orientation for moving toward becoming a cultural warrior for social change is sociotransformative constructivism (STC). Elsewhere (Rodriguez 1998b) I suggest how STC is an orientation that draws from multicultural education (as a theory of social justice) and social constructivism (as a theory of learning). One way to look at STC is that as feminist post-structural frameworks seek to transform oppressive contexts in terms of gender, STC seeks to transform oppressive contexts in terms of teaching and learning in the classroom.

In the following section, I provide an example of how I have used STC in my methods class in an effort to do more than being a cultural worker. My hope is that this example will show how being a cultural warrior requires taking risks—including the risks of examining our privileged power positions as researchers. STC encourages the researcher/educator to gaze inward and to examine the power we hold over whether our research and our teaching are used to transform (or sustain) the pervasive social inequalities in today's schools.

The Courage to Hold the Other's Gaze and Using STC

Sociotransformative work is often labeled "new age tripe" or "touchy feely" by those who fail to understand the courage needed to take risks associated with working against the grain. A growing body of research in multicultural education is making clear how demanding and emotionally draining sociotransformative work really is due to the energy it takes to manage the racist and/or sexist hostility wielded by those who feel threatened by social change (Chavez-Chavez and O'Donnell 1998; Grant 1999). Therefore, whether you are male or female, Anglo or Latino; whatever your ethnicity or sexual orientation, if you are working for social justice, you will encounter those who will seek to silence you, to "place" you, and to intimidate you.

Using my previous work and the work of others, I have grouped the various forms of resistance to social change and to educational reform reported in the literature into two broad areas. These are: resistance to pedagogical change and resistance to ideological change. Resistance to pedagogical change is defined as the resistance to changing one's perceptions of what constitutes being an effective teacher, and it may be influenced by what Lortie (1975) calls an "apprenticeship of observation." That is, student teachers' (and teachers') deeply ingrained notions of what a teacher's job is like from having being a student for 15 or more years. Therefore, if teachers—as students—were mostly exposed to transmissive and teacher-centered education, they are more likely to replicate this approach in their own classrooms.

Resistance to ideological change is defined as resistance to changing one's values and belief systems. Hence, if social justice issues such as, teaching for diversity, gender equality, fighting homophobia, and so on, are "topics" that pre- and in-service teachers only begin to deconstruct during their teacher education program, it is no wonder that this process is going to be long and difficult. Being a cultural warrior, then, involves acknowledging that resistance to teach for diversity and for understanding are part of the rocky terrain of doing sociotransformative work—and the very reasons to navigate through it.

In the article on sociotransformative constructivism (STC) previously mentioned (Rodriguez 1998b), I expand on these two forms of resistance, and on how I used STC to help preservice teachers learn to teach for diversity and understanding. The details of this year-long study are provided in the paper, so I what would like to do here is to only highlight the general features of STC as an alternative orientation for those interested in teaching for diversity and understanding.

Sociotransformative Constructivism
STC is an orientation to teaching and learning based on the notion that knowledge is socially constructed and mediated by cultural, historical, and institutional contexts. Therefore, knowledge production and reproduction

is not perceived as a neutral and mechanical process, but as a complex process by which certain social norms and privileged positions can be maintained or challenged. For example, a student may not just learn physics content in an advanced physics class, but he or she may also learn: cultural codes (e.g., males are usually the ones who take this subject, and the curriculum is usually reflecting an eurocentric male tradition); historical codes (e.g., women and minorities are historically not represented in proportional numbers in advanced classes); and institutional codes (tracking is used as a tool to prevent women and minorities to participate in these courses, the traditional, content-laden exams favor those with abilities to memorize information unrelated to real life situations). A sociotransformative constructivist orientation seeks to engage students in (de)constructing the structures of power from which established cultural, historical, and institutional codes spring. Power, then is a central construct in STC—power is the currency of social change. In this way, the STC orientation provides spaces where the curriculum becomes a site for learning the subject matter in socially relevant and empowering ways—a site where existing contexts can be collaboratively transformed to meet social justice goals.

Given the various forms of resistance discussed earlier, how can teacher educators/teachers implement STC? STC is formed by four closely linked elements: *the dialogic conversation, reflexivity, authentic activity,* and *metacognition.* While it is artificial to discuss each one of these elements separately, it is important to note that they are not listed here in any special order, nor must they follow any sequential pattern to be implemented. It is also important to note that STC is not being proposed as the "theory of everything." My hope is that STC will add to the repertoire of ideas and strategies of those who believe that working for social justice is an imperative today, and that it starts where our everyday practice begins.[4]

The Dialogic Conversation
The dialogic conversation is based on Bakhtin's notion of *dialogicality* (Bakhtin 1981). Voloshinov (cited in Wertsch 1991) explains how dialogicality means more than the engaged dialogue of speaking voices: "To understand another person's utterance [unit of speech] means to orient oneself with respect to it, to find the proper place for it in the corresponding context . . . Any true understanding is dialogic in nature" (54). Therefore, in order for me (a Latino) to have a meaningful and dialogic conversation with, for example, a middle-class, Anglo-male, phys-

ics, preservice teacher, who opposes preferential treatment of underserved students to get into college, we would both need not only to listen to our words (utterances), but to understand (and respect) how each other's different voices (or "speaking consciousness," Bakhtin 1981) reflect our different contexts and experiences. In this way, classroom discussions move beyond merely understanding what is being said to understanding the reasons why the speaker chooses to say what he or she says in a particular context. The dialogic conversation allows the listener/speaker to ask the ultimate Bakhtinian question, "who is doing the talking"? In other words, whose voice (insights, values, and beliefs) are being represented by the speaker/listener?

As one could imagine, to expose one's voice in the presence of others with different power hierarchical positions can be problematic. In fact, Bakhtin has been criticized for not considering how dialogue can be coercive or threatening (Morris 1994). Therefore, for the dialogic conversation to take place in the classroom, the establishment of trust is essential.

An example: During the first three weeks of my methods class, students are given current statistical information on pupils' achievement in science by socioeconomic status and gender within ethnic groups. Students are also engaged in a critical study of the nature of science through which they are asked to reflect on questions such as: Whose interests are being served by science and scientists? Whose science is traditionally taught in our schools?

Analysis of the preservice teachers' resistance to ideological change shows that it could be interpreted in relation to three underlying themes or types of reactions to the class activities and discussions: feelings of guilt, shame, and defensiveness; feelings of hopelessness; and feelings of being overwhelmed (Rodriguez 1998b). I will explain only the first theme here. Feelings of *guilt, shame,* and *defensiveness* have to do with how some preservice teachers react when exposed to information about the current inequities in our schools. Some students resisted confronting these feelings by using an avoidance strategy. For example, some exclaimed in class, "Why do we have to talk about multicultural education in this class? This should be covered in other classes," or "There is too much emphasis on multicultural education in this class. Why do I have to learn about multicultural education if I want to teach in rural schools?" Other preservice teachers demonstrate a more aggressive form of resistance to ideological change by posing issues such as: "I don't see what the fuss is all about; if women and minorities want to do well in science, they are just going to have to work as hard as White males!" This type avoidance strategy whereby racism, discrimination, and other social inequities are talked about as if they are "somebody else's concern" have been reported by other researchers in a variety of contexts (Macintosh 1989; Cochran-Smith 1991, 1995; Ahlquist 1991; King 1991; Sleeter 1994; and others). Sonia Nieto reiterates that "some of these concerns may be sincere expressions of protectiveness toward students, but others are merely self-serving and manifest teachers' discomfort with discussing racism" (1996, 89).

Although it is difficult to listen to the students' comments against equity and social justice in the methods class, I find comfort in the fact that after a few weeks of classes we had established a classroom environment that at least allowed for honest and open interactions with me and among their peers (dialogic conversation). It is at this time that some of the methods students finally begin to seriously articulate their beliefs and perceptions about people unlike themselves, and how their perceptions might come to bear on their teaching practice. It is at this point that the assigned readings and classroom activities become "wake up calls," we begin to move away from the realm of well-intended education reform policies and away from the well-intended goals of our teacher education program to discussing directly how (or whether) equity could be enacted in their future science classrooms. The types of personal and social transformations some students begin to consider can be better appreciated through the process of reflexivity explained below.

Reflexivity

Reflexivity involves becoming aware of how one's own social location (e.g., ethnic and cultural backgrounds and socioeconomic status), ideological location (e.g., belief systems and values), and academic location (e.g., education and skills) influence what we consider important to learn. Therefore, through reflexivity the STC orientation not only involves student-centered, hands-on, minds-on activities aimed at learning science content, but also involves a discussion of how science knowledge is produced and reproduced, who are (were) recognized as scientists, how their work influences society at large, and how social issues determine which scientific work is worth funding. Therefore, reflexivity opens a window for students and teachers to examine the culture of power and explore ways to transform it for the benefit of all and not just the privileged few. Needless to say, exploring one's own position of power (or lack of it) is a difficult and often uncomfortable process, but as Jay Lemke (1995) explains, reflexivity can give us the tools to better understand ourselves and our actions (or inactions). Lemke states that "It is our praxis [our theories in action] that must be self-reflexive; we must analyze and criticize our own process of analyzing and criticism, and all our newly created practices of every kind for getting inside of, and outside of, what we and other people do to make the kinds of meanings we make" (1995, 157)

An example. In my methods class, when we have began to establish an environment of trust, where dialogic conversations are possible, one way to encourage students to come face to face with their taken for granted assumptions is requiring them to read and write reactions to (not summaries of) articles by Joyce King (1991) on dysconscious racism, by Christine Sleeter (1994) on white racism, and chapters in Michael Reiss' (1993) book, *Science Education for a Pluralistic Society*. Three chapters in the latter describe many contributions to the physical and biological sciences by peoples around the world and throughout history. The conversation that ensued about these readings is usually very intense, emotional, and revealing. For example, a female chemistry preservice teacher explained her reaction to Sleeter's (1994) article, "White racism," as follows: "I do think that often white people are afraid to voice opinions about issues of diversity for fear of being labeled a racist. I agree with the author in that often whites are silent about racism, pretending that it does not exist . . . .Often white people plead ignorance with regards to such issues, and it becomes the people's of color responsibility to educate everyone else. This is not fair, but I have to admit that I'm guilty myself" (Rodriguez 1998b, 605)

This type of difficult, yet necessary, introspection is an example of how reflexivity in the sociotransformative constructivist orientation can help preservice teachers explore their positions of privilege. In this way, they are encouraged to begin moving from a stance of avoidance and selective ignorance to a stance of self-awareness and transformative action. This is not an easy task because the same strategies of counter-resistance (having students read and react to key articles in writing and during group discussions) may intensify the feelings of discomfort and apprehension toward working with pupils from diverse backgrounds. This is one of the reasons preservice teachers must be provided with authentic activities to be able to connect ideological changes (e.g., all students can succeed in science) with pedagogical strategies that facilitate that process. This topic will be discussed in more detail in the next section, but I must first point out that the term counter-resistance is chosen here to highlight the importance of being a cultural warrior. In other words, If one of the goals of the methods class (and of a teacher education program) is to prepare teachers to teach for diversity and understanding, and we know that these goals will be met with resistance—resistance that will not be very different from that encountered by teachers who teach against the grain in public schools—then we must be forthcoming with our political agenda. This is not to say that students should be penalized with lower grades if they disagree with the teacher education program's ideological and epistemological commitments, but students deserve to know that different teacher education programs have different expectations, and different ways of assessing whether those expectations are being met. In my methods class, for instance, homogeneous and instantaneous ideological agreement was neither sought nor expected. The main objective was to help students engage with me and with each other in fruitful dialogical and reflexive conversations which could open opportunities for all of us to grow as effective teachers.

Metacognition

Another important element of STC is *metacognition*. This construct is primarily borrowed from individual constructivist research. Baird (1990, cited in Gunstone, 1994, 134) defines metacognition as "the knowledge, awareness and control of one's own learning." Gunstone (1994) adds that learners can be instructed to be metacognitive by encouraging them to ask themselves questions such as, What am I meant to be doing? Do I know what to write/look for? What is the purpose of this task? Have I done everything necessary? Can I explain this to someone else? These are indeed important questions, but STC expands the concept of metacognition to include a deeper form of critical engagement. By deeper and more critical I do not just mean in the sense of developing higher order learning skills. In this case, I am referring more to the process of developing a sense of consciousness and agency on one's own ways of learning (Freire 1989).

In other words, it is not enough to encourage students to reflect on how they learn, without encouraging them to reflect on the what and why they are being asked to learn. Therefore, a learner in the STC classroom may in addition ask Why am I learning this? Why am I doing it this way? What control do I have in how to proceed? By what other method can I learn this best?

An example. In my methods class, students' resistance to pedagogical change is usually manifested in two general forms. At the personal level, one of these forms of resistance is rooted in their perceptions of sociotransformative constructivism as an unviable alternative to the transmissive model of learning which worked (and still is working) well for them during 15 years of their formal schooling. The other form of resistance appear to be more extrinsic, and had to do with what the preservice teachers perceive as the shortcomings of sociotransformative constructivism to realistically meet the many demands of teaching (i.e., cover the content-laden curriculum, satisfy parents' expectations, prepare pupils for college, and so on.).

Some preservice teachers argue that the collaborative learning focus of social constructivism is not useful for all kids. Consider this example from a male physics student, "I disagree with the blanket idea that cooperative learning is the best way to learn science. My own experience has been that learning on my own was better when it came to: problem solving; academic achievement; preference—I usually prefer individual work, except for laboratories" (Rodriguez 1998b, 610). As a strategy of counter-resistance in these types of situation, students are asked to engage in collaborative learning activities in interdisciplinary groups throughout the course. In addition, they are provided with opportunities to deconstruct the value of the activities through group discussions and selective readings. For instance, this is how another student—a female, biology preservice teacher—reflected on her experiences as a sole learner after reading selected chapters from Robertta Barba's text, *Science in the Multicultural Classroom: A Guide to Teaching and Learning*: "I remember most of my learning took place as a solitary activity, mainly listening, reading or memorizing. Even in my science classes I do not remember doing activities in a group. Reading over the benefits of cooperative learning made me realize how much I was cheated out of a good education" (Rodriguez 1998b, 611).

Through dialogic conversations in class with me and their peers, reflexive writings and reactions to assigned readings, and authentic activities (e.g., working in collaborative groups to discuss the benefits and shortcomings of collaborative learning), some preservice teachers began to become more metacognitive and critique how their past and present histories as learners were influencing their emerging identities as teachers.

The second form of resistance to pedagogical change involved aspects of social constructivism which, in the preservice teachers' point of view, made this orientation simply unworkable in "real" schools. For instance, after asking students to read key articles on performance/authentic assessment, they were required to discuss how this form of assessment was tied to the elements of the sociotransformative constructivist orientation (STC). Several students argued that implementing this form of assessment and using STC would take too long to cover the prescribed curriculum. Others found performance/authentic assessment to be "too subjective, to be fair to kids." Even one student who found performance assessment in science very appealing felt frustrated by the amount of work that it would take to prepare meaningful performance assessment tasks. She added, "I have been in college for almost four years, and I am going to have to relearn everything before I teach it" (Rodriguez 1998b, 611). All of these constraints mentioned by the students are of course quite valid and very difficult for any beginning teacher to change by himself or herself. It is at this intersection where the differences between the teacher educator as cultural worker and the teacher educator as cultural warrior are defined. For the cultural warrior these moments of dissonance between the teacher education program's goals (prepare teachers to teach for diversity and understanding) and the institutional, social, and economic realities of today's schools provide excellent points to engage not only in intellectually honest dialogic conversations, but on specific demonstrations of how to effect change. In this way, these points of dissonance will not end up leading to feelings of despair and hopelessness commonly reported in the literature. On the contrary, acting on points of dissonance can lead to empowering moments of possibilities where teacher educator and students explore the multiple roles even an individual teacher can play to effect change in demanding school contexts.

Perhaps the easiest way to appreciate how the various elements of STC can be enacted is through investigating what it is meant by authentic activity in this orientation.

Authentic Activity

One essential aspect of being a cultural warrior for social change in teacher education is practicing what one preaches ideologically as well as pedagogically. Therefore, many authentic opportunities must be provided for students to experience, critique, and reflect on the differences between traditional ways of teaching and alternate orientations. For example, one of the differences between the STC approach and others proposed under individual constructivism is that in addition to doing minds-on, hands-on activities students are urged to reflect on how the subject under study is socioculturally relevant and tied to everyday life. Furthermore, STC strongly cautions that hands-on activities may not necessarily be minds-on, or may even have little to do with science or society at large. As John Dewey eloquently points out, "To believe that all genuine education comes through experience does not mean that all experiences are genuinely or equally educative" (1938, 25).

In fact, in many of my visits to schools or attendance at teacher conferences I have observed how student-centered, constructivist learning is often translated into a busy array of teacher-guided activities that have little to do with learning science and even less with learning for understanding. For instance, I once attended a workshop that carefully described how students could be challenged to construct a balloon-popping machine. The design was similar to that found in the mouse trap construction kit sold at toy stores. The presenters provided detailed guidelines for how to organize the students, where to get the materials, and how to conduct the activity for over two-weeks worth of classes. Even though this activity was no doubt fun and hands-on, I argued at the workshop that it had little to do with science. The presenters never mentioned that students should be required to: formulate a hypothesis; consider the variables in their "experiments;" make connections to engineering and the work engineers do; modify the prescribed design to improve it or address their own ideas; make connections to mathematics; nor were students encouraged to reflect on how doing this activity was socially relevant and related to every-day life. Hedy Moscovici and Tamara Nelson (1998) call this type of phenomenon "activitymania," and they provide useful suggestions for how a teacher or teacher educator can move from this busy approach to a more inquiry-based, student-centered, and socially relevant orientation.

Learners (including pupils, pre- and inservice teachers, and science teacher educators) must be provided with opportunities to engage in activities that closely resemble those commonly carried out by practitioners in the community of practice of the subject under study. This notion is congruent with current science education reform initiatives, and I suggest this is one place where students' diverse backgrounds and abilities can also be used as tools to enhance learning by allowing them to contribute their expertise and ideas through role-playing, group research projects, and so on.

An example. In previously published work (Rodriguez 1998b), a detailed description of an authentic activity used with secondary science methods students is provided to illustrate the various components of STC. This activity also shows how various individual constructivist pedagogical strategies, such as concept mapping; predict, observe, and explain (POE); translational activities, and so on (Baird and Mitchell 1986; Mitchell and Mitchell 1992, have adapted to fit with the STC orientation. This time, an activity used in my elementary science methods course will be described to illustrate how activitymania can be fun and hands-on, but provides few opportunities for learning for diversity and understanding.

I begin by telling students to get into groups of three's and by distributing copies of the activity sheet. Students are told that we are going to learn how to make ice-cream and that we will have a chance to eat it too. One student from each group is selected to gather the simple materials needed (zip lock bags, salt, vanilla, milk, pieces of fruit, spoon, measuring cups, and ice). For the next half-hour, students enthusiastically follow the written recipe for making ice-cream. By the noise level in the room, anyone can tell that students are actively engaged with one another, and by the way they enjoyed eating the ice-cream, anyone can tell that this is one activity they would most probably implement in their own classrooms.

An interesting dialogic conversation follows when I ask students to label the type of pedagogical activity we just tried according to the theories of learning we have previously covered. Most usually give similar responses, "It was an example of discovery learning;" "It was a fun constructivist idea;" "It was a social constructivist strategy because we were using everyday things to do science." A few other students usually catch on and state that they had not learn any scientific concepts, "we have not covered any science yet;" "Are we doing anything else after making the ice-cream?;" "This was just fun, but we have not learned anything." These types of comments are exactly the type of issues a sociotransformative teacher educator/teacher would like to see his/her students raise when exposed to an example of activitymania.

The next half-hour, dialogic conversation provides student teachers with a concrete example to reflect on what doing authentic science activities in the classroom is all about. By making clear that following a recipe to make ice-cream in collaborative groups has little to do with critically exploring science, students begin to reflect more carefully on what aspects of the lesson were needed to make it more socially constructivist. For example, students usually point out that pupils should be given a chance to figure out the scientific concept behind this process (the freezing point of water is lowered by adding salt to the ice). Some suggest that the activity can be done first, and then have students investigate the factors responsible for the observed phenomenon (e.g. provide thermometers; allow them to change some of the elements in the recipe or variables). Other students have suggested that the activity can be made more socially relevant by making ties to how salt is used on icy roads during the winter, or by having students investigate who invented ice-cream and for how long this invention has been around. Other suggest to invent different kinds of ice-cream for people who are lactose intolerant, diabetic, or who wish to reduce weight.

By having students critique an example of an activitymania lesson, they can arrive to a better understanding that just fun and collaborative activities and/or magic demonstrations have no room in the sociotransformative science classroom. Doing authentic science in the classroom means assisting students to critically explore the accepted scientific views in relation to their everyday lives.

In short, I have illustrated here how sociotransformative constructivism suggests that to become effective multicultural science teachers, preservice teachers need to also explore how their own belief and value systems intercept with their theories of learning. As preservice teachers in collaboration with the instructor begin to understand how these interceptions facilitate or obstruct their progress toward becoming effective teachers or cultural warriors, steps are taken to continue moving toward socially relevant professional development and transformative action.

The Courage to Be(come) Cultural Warriors

We must dare in order to say scientifically that we study, we learn, we teach, we know with our entire body. We do all these things with feeling, with emotion, with wishes, with fear, with doubts, with passion, and also with critical reasoning...We must dare as never to dichotomize cognition and emotion...We must dare so that we can continue to do so even when it is so much more materially advantageous to stop daring (Freire 1998, 3).

Using the words of master teacher, Paulo Freire, to guide me, a summary of the arguments and recommendations I have made so far is presented in three general areas: Using multiple theoretical frameworks to inform our work; Expanding our research agenda; and Agency. Each one of these is elaborated below.

Using Multiple Theoretical Frameworks to Inform our Work

It was argued that individual constructivism has been the dominant theoretical framework guiding most studies in science education for over twenty years. Although science education research based on this orientation has been very fruitful, it appears to have mainly benefited the science education research community. That is, to date, there is little evidence that findings from this body of research have had an impact on: How teachers continue to teach; how new teachers learn to teach; how students learn; and most importantly on the pervasive gaps in student achievement and participation among traditionally underserved children and children from Anglo and Asian[5] ethnic backgrounds.

The continuing and escalating reproduction of social inequalities in schools demand that we, as a research community, gaze inward and explore our role in sustaining those inequalities by virtue of the work we choose to do or not to do as privileged intellectuals in the culture of power. It was suggested that by using theoretical frameworks such feminist poststructuralism and/or sociotransformative constructivism—for example—we could begin a process of making our research more socially relevant and responsive to the realities of today's schools.

Expanding our Research Agenda

It was also suggested that a substantial body of research from the multicultural education and learning to teach literature support the notion that pre- and inservice teachers displayed two major types of resistance to change. Resistance to pedagogical change (i.e., resistance to changing one's perceptions of what constitutes being an effective teacher) and resistance to ideological change (i.e., resistance to changing one's values and belief systems). What we now need is to acknowledge that these types of resistance exist, that they are embedded in the process of change, and to begin exploring effective strategies of counter-resistance. That is strategies that provide meaningful examples of how teachers can carry out the often complex and risky changes we expect them to implement in their classrooms. This also means that we, as researchers and teacher educators, must work in the trenches with teachers and provide them with the onsite and continuing support they need to see our (their) proposed changes through.

Agency

Agency refers to the conscious role we choose to play in helping bring about social change for the collective benefit of all, especially those in lower hierarchical power or disadvantaged positions than ourselves. This involves moving away from the notion of suppressed subjectivities. That is, the modern and positivistic illusion that for our work to be scientific, it must be objective. If we follow Freire's advice (in the above quote) and if we dare not to separate cognition and emotion, then we will embrace our ethical responsibilities and ensure that the researched also benefits from our research. We often see a form of *truncated agency* even in studies that claim to give voice, or to empower the other. We must distinguish between agency and truncated agency in terms of how the researcher participants benefited from participating in the study. Doing research on the Other, providing robust analyses of their disadvantaged positions, and using strong theoretical frameworks to support our claims, does little to change the very oppressive forces we eloquently describe. This is of course unless we embed a form of *catalytic validity* into our research. That is, a way of conceptualizing our research as valid by the degree to which participants and researchers have substantially improved their condition as a direct result of their involvement in the study. This also implies the long-term benefits of having acquired new insights and understandings of one's own abilities for transformative action.

To close, I must return to the original question, how do we help improve the preparation of our teachers and the achievement and participation of all students in science? My belief is that what we really need is courage. We have accumulated a great body of knowledge and ideas about learning, about teaching, and about the social issues that prevent students from learning, and teachers from teaching. What we need then is to muster the courage to manage the resistance and the risks associated with helping implement our proposed changes. This will take the collective effort of our research community, but it must start at the individual level. To have the courage to escape the entrapment of our own privilege is to take a step closer to understanding how our actions and inactions make us either pawns for the status quo or effective agents for transformative social justice. We must have the courage to expand our gaze as researchers and teacher educators and rise to the challenge of becoming cultural warriors for social change.

References

Ahlquist, R. 1991. Position and imposition: Powser relations in a multicultural foundations course. *Journal of Negro Education.* 60: 158–160.

Baird, J. R., & Mitchell, I. J. 1986. Improving the quality of teaching and learning: An Australian case study—the PEEL project. Melbourne, Australia: Monash University,

Bakhtin, M. M. 1981. The dialogic imagination. Austin, TX: University of Texas Press.

Barba, R. 1995. Science in the multicultural classroom: A guide to teaching & learning. Nedham Heights, MA: Allyn & Bacon.

Carter, K. 1990. Teachers' knowledge and learning to teach. In W. R. Houston (Ed.), Handbook of research on teaching education (pp. 291–310). New York: McMillam.

Chavez-Chavez, R., & O'Donnell, J. 1998. Speaking the unpleasant: The politics of (non) engagement in the multicultural education terrain. New York: State University of New York Press.

Cochran-Smith, M. 1991. Learning to teach against the grain. Harvard Education Review, 61: 279–310.

Cochran-Smith, M. 1995. Color blindness and basket making are not the answers: Confronting the dilemmas of race, culture, and language diversity in teacher education. American Educational Research Journal, 32: 493–522.

Day, J. C. 1993. Population projections of the United States, by age, sex, race, and Hispanic origin: 1993 to 2050. (P25-1104). U.S. Bureau of the Census.

Dewey, J. 1938. Experience and education . New York: Macmillan Publishing Co.

Fensham, P., Gunstone, R., & White, R. 1994. The content of science: A constructivist approach to its teaching and learning. New York, NY: Falmer Press.

Freire, P. 1989. Pedagogy of the Oppressed (31 ed.). New York: Continuum.

Freire, P. 1998. Teachers as cultural workers: Letters to those who dare teach. Boulder: Westview Press.

Gabel, D. 1994. Handbook of Research on Science Teaching and Learning. Toronto: McMillam Publishing Company.

Grant, C. 1999. Multicultural research: A reflective engagement with race, class, gender, and sexual orientation. London: Falmer Press.

Gunstone, R. F. 1994. The importance of specific science content in the enhancement of metacognition. In P. Fensham, R. Gunstone, & R. White (Ed.), The content of science: A constructivist approach to its teaching and learning. New York, NY: Falmer Press.

Kagan, D. 1992. Professional growth among preservice and beginning teachers. Review of Educational Research, 62: 129–169.

King, J. 1991. Dysconscious racism: Ideology, identity, and the miseducation of teachers. The Journal of Negro Education, 60: 133–146.

Lather, P. 1991. Getting smart: Feminist research and pedagogy with/in the postmodern. New York: Routledge.

Lee, S. J. 1996. Unraveling the "model minority" stereotype: Listening to Asian American youth. NY: Teachers College Press.

Lemke, J. 1995. Textual Politics: Discourse and social dynamics. Bristol: Taylor & Francis.

Lortie, D. 1975. School teacher. Chicago: University of Chicago Press.

Matthews, M. R. 1994. Science teaching: The role of history and philosophy of science. New York: Routledge.

McMillen, M. M., & Kaufman, P. a. 1994. Dropout rates in the united states: 1994 (NCES 96-863). Washington, DC: National Center for Educational Statistics.

MacIntosh, P. M. 1989. White privilege: Unpacking the invisible knapsack. *Peace and Freedom*, 10–12.

Mitchell, J. and Mitchell, I. 1992. Some classroom procedures. In J. Baird and J. R. Northfield (Eds.), Learning from the Peel experience. Melbourne, Australia: Monash University Printing Services.

Morris, P. 1994. The Bakhtin reader: selected writings of Bakhtin, Medvedev, Voloshinov. London, UK: Edward Arnold.

Moscovici, H., & Nelson, T. 1998. Shifting from activitymania to inquiry. Science and Children, (January), 14–17.

National Science Foundation (NSF). 1996. Indicators of Science and Mathematics Education 1995. Arlington: NSF.

Nieto, S. 1996. Lessons from students on creating a chance to dream. In H. E. Review (Ed.), Working together toward reform (pp. 77–113). Cambridge, MA: Harvard Educational Review.

Oakes, J., & Lipton, M. 1998. Teaching to change the world. Boston: McGraw-Hill College.

Oakes, J., Ormseth, R., & Campbell, P. 1990. Multiplying inequalities: The effects of race, social class, and tracking opportunities to learn mathematics and science (NSF). Santa Monica, CA: RAND.

Orner, M. 1992. Interrupting the calls for student voice in "liberatory" education: A feminist poststructuralist perspective. In J. Gore, & C. Luke (Ed.), Feminisms and critical pedagogy. New York: Routledge.

Peng, S. S., & Hill, S. T. 1995. Understanding racial-ethnic differences in secondary school science and mathematics achievement (NCES 95-710). Washington, DC: National Center for Education Statistics.

Pfundt, H., & Duit, R. 1991. Bibliography: Students' alternative frameworks and science. Kiel, Germany: IPN.

Reinharz, S. 1992. Feminist methods in social research. New York: Oxford University Press.

Reiss, M. 1993. Science education for a pluralistic society. London, UK: Open University Press.

Rodriguez, A.J. 2000. Linking Bakhtin with feminist poststructuralism to unravel the allure of auto/biographies. Research in Science Education, 30, 13–22.

Rodriguez, A. J. 1998a. Busting open the meritocracy myth: Rethinking equity and student achievement in science. Journal of Women and Minorities in Science and Engineering, 4(2&3): 195–216.

Rodriguez, A. J. 1998b. Strategies for counterresistance: Toward sociotransformative constructivism and learning to teach science for diversity and for understanding. Journal of Research in Science Teaching, 36(6): 589–622.

Romo, H. D. and T. Falbo. 1996. Latino high school graduation: Defying the odds. Austin: University of Texas Press.

Schmidt, W. 1998. A summary of Facing the Consequences: Using TIMSS for a closer look at United States mathematics and science education. Ann Arbor: Third International Mathematics and Science Study:

Secada, W., R. Chavez-Chavez, E. Garcia, C. Muñoz, J. Oakes, I. Santiago-Santiago, and R. Slavin. 1998. No more excuses: The final report of the Hispanic dropout project. Washington: U.S. Department of Education.

Sleeter, C. (1994). White racism. Multicultural Education, 1(4), 5–8,39.

Stallings, J. A. 1995. Ensuring teaching and learning in the 21st century. Educational Researcher, 24(6): 4–8.

Wertsch, J. V. 1991. Voices of the minds a sociocultural approach to mediated action. Cambridge, MA: Harvard University Press.

Zeichner, K. M. and J. M. Gore, J. M. 1990. Teacher socialization. In W. R. Houston (Ed.), Handbook of research on teaching education (pp. 329–348). New York: Macmillan.

Notes

1. In an article for Research in Science Education, (Rodriguez, 2000, in press), I use Bakhtin and feminist poststructuralist orientations to deconstruct the allure of (auto)biographical accounts.

2. Traditionally underrepresented students in science are women and students from Latina/o, African, First Nations (Native Americans), and from some Asian ethnic groups.

3. Individual constructivism is so widely used in science education research that the word *constructivism* has become a blanket term in the literature. For more a detailed discussion of the various "faces" of constructivism see Rodriguez (1998b).

4. I am currently conducting several research projects with teachers, preservice teachers, and their students using the STC orientation. These studies shall provide more insights into whether STC can have the kind of transformative impact it hopes to create.

5. Following Lee's (1996) suggestion the term "Asian" is used here with caution to avoid the "model minority" stereotype. There is more cultural, socioeconomic, and academic performance diversity among peoples grouped under this general ethnic term, but these are rarely reported.

List of Contributors

Angela Calabrese Barton, Editor, University of Texas and Teachers College Columbia University

Angela Calabrese Barton is an associate professor of science education and the Director of the Urban Science Education Center at Teachers College Columbia University. Her research interests include feminist and critical perspectives in science education and its implications in poor urban settings. Some of her publications appear in Curriculum Inquiry, Journal of Research in Science Teaching, Journal of Teacher Education, and Women's Studies Quarterly. She has been the recipient of the National Association for Research in Science Teaching Early Career Award and the National Science Foundation Early Career Award.

Julie A. Bianchini, University of California, Santa Barbara

Julie A. Bianchini is an assistant professor of science education. Her research interests include research into the nature of science and the work of scientists and its implications in secondary education and teacher education. Some of her publications appear in Journal of Research in Science Teaching and Science Education.

David J. Brady, University of Illinois at Urbana-Champaign

David J. Brady is an associate professor of electrical and computer engineering at Duke University.

Heidi Bulmahn Barker, University of Illinois at Urbana-Champaign

Heidi Buhlman Barker has a Ph.D. in Curriculum and Instruction from the University of Illinois and is now living in Denver Colorado.

Tanahia Burkett, University of Texas at Austin

Tanahia Burkett completed her Ed.M. degree at The University of Texas at Austin and is currently teaching elementary school in New Orleans

Parish, New Orleans, LA. Her research and teaching interests include empowering education in poor, urban settings.

Lynette M. Cavazos, University of California, Santa Barbara

Lynette M. Cavazos is Coordinator of Secondary Teaching Credential Program and Science Partnership for Schools. Her research areas and interests include gender and equity issues in science education and the professional development of beginning and experienced teachers. Her publications can be found in the Journal of Research in Science Teaching.

Dana Fusco, York College, City University of New York

Dana Fusco is an assistant professor of educational psychology. Her research interests include authentic assessments, critical theory, and student action research. She has several publications in the Journal of Research in Science Teaching.

Gaell M. Hildebrand, University of Melbourne, Australia

Gaell M. Hildebrand is a professor in Classroom Teaching and Learning. Her research interests include writing and feminist and poststructural perspectives in science education and its implications in teachers' constructions of science and school practice. Some of her publications appear in the Journal of Research in Science Teaching.

Elaine Howes, Teachers College, Columbia University

Elaine Howes is an assistant professor of science education. Her research interests include research into elementary science teacher education and feminist theory. Some of her publications appear in Journal of Research in Science Teaching.

Margery D. Osborne, Editor, University of Illinois at Champaign-Urbana

Margery D. Osborne is an associate professor of science education. Her research interests include feminist and poststructural perspectives in science education and its implications in early elementary settings. Some of her publications appear in Journal of Research in Science Teaching, Journal of Curriculum Studies, and Women's Studies Quarterly.

Sharon Parsons, San Jose State University

Sharon Parsons is an associate professor of science education. Her research interests include feminist and poststructural perspectives in science education and its implications in school university collaborations.

Some of her publications appear in the International Journal of Science Education and Research in Science Education.

Alberto J. Rodriguez, New Mexico State University

Alberto J. Rodriguez is an associate professor of science education. His research interests include critical and multicultural perspectives in science education. Some of his publications appear in Journal of Research in Science Teaching and the Journal of Teacher Education.

Kathleen St. Louis, University of Texas at Austin

Kathleen St. Louis is a doctoral student at The University of Texas at Austin. Her research and teaching interests include empowering education in poor, urban settings.

Matthew Weinstein, Macalester College

Matthew Weinstein is an assistant professor at Macalaster College, St. Paul, Minnesota. His research interests focus on cultural studies and science education. His publications have appeared in Educational Policy. His book, Robot World, was published with Peter Lang Press.

Michalinos Zembylas, University of Illinois at Urbana-Champaign

Michalinos Zembylas has a Ph.D. in science education from the University of Illinois and is now teaching at Michigan State University.

Index

Agency, children 38; consciousness and 338; discourse of 23; embodied 18; race, class culture 23; possibilities of 248-9; truncated 343-345
American Association for the Advancement of Science, 10-14, 19, 90, 153, 202, 216-217
Anti-Colonialist, 165
Apple, Michael, xii, 12
Assessment, authentic 216-218; rethinking the nature of 220-222
Assimilation, 12
Atwater, Mary, 19, 20-21, 263
Authority, challenges of 188-190; knowledge 24; medical 243; of others 114; scientific 136-41
Autobiography, as research 81-85, 295, 311-315, in teaching 133-134

Bachelard, Gaston, 53
Bacon, Francis, 179-180
Ball, Deborah 263-265
Banks, James, 84, 265
Barton, Angela Calabrese, 132, 203, 217, 243, 263, 296
Belmont Report, 234-235
Betelheim, Bruno, 38, 50
Boler, Megan, 122

California Science Content Standards, 262

Cisneros, Sandra, 59, 61
Class, 1, 12-15, 23-25, 84, 134-8, 238, 306, 310
Code, Lorraine, 108
Coelho, Paulo, 38, 50
Coles, Robert, 38
Consciousness-raising, 4, 133-138
Conservatism, 109
Cultural warrior, 325, 343-345
Curriculum, 2, 4, 5, 14, 25, 49, 61-78, 105-110, 131, 146, 162, 212, 260-270, 325

D'Haen, Teresa, 46
Darwin, Charles, 181
DeBoer, George, 106
Dewey, John, 71, 340
Dialogicality, 334
Domination, nature 17-18; power and 1, 23-25; western 222
Duckworth, Elizabeth, 71

Einstein, Albert, 141, 181
Eisenhart, Margaret, 19, 21, 203, 208
Ellsworth, Elizabeth, 46
Emotions, 99-100; emotional genealogies 114-116, 120-123; emotional understanding 112-113; as en entity 111; in teaching 102; transforming 104
Empowerment, 3, 21, 23-25, 264, 305-311
Epstein, Stephan, 16

Fausto-Sterling, Anne, 271
Fee, Elizabeth, 136
Feminism, and Science 13-20
Feminist pedagogy, emancipatory possibilities 13-25, 295-324; patriarchy, politics and transformation 295-324; postmodernism and 298-325; within the field of science education 13-25, 129-150, 264-268

Foucault, Michael, 19, 107, 115
Freire, Paulo, xii-xiii

Gallas, Karen, 150
Garrison, Jim, 102
Giroux, Henry, xii-xiii
Goals 2000, 10
Gore, Jennifer, 295-299
Greeley, Anthony, 52, 54

Haggerty, Sharon, 165
Haraway, Donna, 16-18, 230-233, 237-243
Harding, Sandra, xvi, 15, 187, 232-234, 236-238
Harre, Rom, 111-112
Hart, Christina, 167
Heidegger, Martin, 38, 44
Helms, Robert, 246
Hollingsworth, Sandra, 63
Homelessness, 79
hooks, bell, 85

Identity, labeling 1; politics and 124, 232, 298, 325; professional 105; race and 82-86; science and 1, 17-25

Journal of Research in Science Teaching, 1-2

Keller, Evenlyn Fox, 17-18, 135
Kuhn, Thomas, 180
Kyle, William, xi

Ladson-Billings, Gloria, 205

Language, and identity 107; and meaning 40; science 17, 80, 166-173; second language learners 21, 276, 311
Latour, Bruno, 109
Layton, David, 167
Liberatory, pedagogies 165, 264-5; science education 4
Life history, 296
Locke, David, 176-178
Loucks-Horsley, Susan, 266-267, 286
Luke, Carmen, 295
Lyotard, Francois, 25, 26

Master texts, 264, 300-313
Mayberry, Maralee, 264-265
McClintock, Barbara, 147
McClintock Collective, 267
McIntosh, Peggy, 26
Middleton, Sue, 296-299
Modernity, 105-106

National Council on Education Standards and Testing, 204
National Research Council, 10, 19, 231, 243-244
National Science Education Standards, 10, 19, 231, 243-244
Nature of science, 14-16, 64, 188, 203, 216, 284, 300, 314, 332-333
Newton, Isaac, 181
Nieto, Sonia, 265
Noffke, Sue, 205

Oppression, economic, 165; empowerment and 306; structures of 231-2, 264; women's 15

Paralogy, 99-100
Politics, assimilation, 13-15; disruption 165-170; feminist 114; guinea pig science 231-242; identity 122, 234
Postmodernism, 25-26; 42, 298-306; defined 306-310
Poverty, Teaching children in, 79-98
Power, discourse 161-165; power/knowledge 12-15, 22, 184-

186, 244, 298; relationships 23-25, 97-106, 132, 232, 295; researcher 63, 242, 264; science 88, 183-190
Pre-service education, 129-155; 268-270; 325-343
Price, Jeremy, 263-265
Professional Development of Teachers, models 263-267
Progressivism, 12, 49

Race and racism, class, gender and 1, 12, 20-25, 82-84, 248-50; race and class 82-84; racism and 331-340; science 165-167; science education reforms 11-12, 230-232
Reflexivity, 336-338
Reiss, Michael, 337
Resistance, 6, 45, 114, 188-190, 299-300, 312-313; student 286-288; to teaching for diversity 326-330
Rosser, Sue, 265
Rowell, Patricia, 168

Sartre, Jean Paul, 53
Schon, Donald, 43
Science, authority 136; the canon of 88-89, 91; creativity 190; emergent 94-96; experiences 35-50, 89-90; inclusive 285-286; nature of science 14-16, 64, 180-181, 188, 203, 216, 284, 300, 314, 332-333; perceptions 139-154; stereotypes 134; transformative 94-96; writing 161-192

Science and Society, 17-19
Science Education through Portfolio Instruction and Assessment, 212
Smith, Dorothy, 232
Social justice, 6, 266-268, 323-325, 331-343
Spillane, 72
Standpoint theory, 232-235, 240-242
Student, lifeworlds 79-81; 94-96; resistance 328; voice 3, 18-22, 146-148, 160-162, 202-204, 216-218

Teachers, identity 105-107; political nature of teaching, 23, voice 169-172
Toulmin, Stephan, 105
Truth, 5, 46-48, 103-115, 123-127, 167, 178-185, 298-299

Urban education, 79-98

Villenas, Sophia, 84
Voice, student 3, 18-22, 146-147, 161, 203, 216-217; teacher 168-170

Weiler, Kathleen, 3, 24
Wittgenstein, Ludwig, 38, 45, 100
Women Educators of Science and Technology (WEST), 259-262, 267-270
Women in Scientific Literacy Project, 259, 270-273
Women writers, 73
Woolf, Virginia, 60, 70

Studies in the Postmodern Theory of Education

General Editors
Joe L. Kincheloe & Shirley R. Steinberg

Counterpoints publishes the most compelling and imaginative books being written in education today. Grounded on the theoretical advances in criticalism, feminism, and postmodernism in the last two decades of the twentieth century, Counterpoints engages the meaning of these innovations in various forms of educational expression. Committed to the proposition that theoretical literature should be accessible to a variety of audiences, the series insists that its authors avoid esoteric and jargonistic languages that transform educational scholarship into an elite discourse for the initiated. Scholarly work matters only to the degree it affects consciousness and practice at multiple sites. Counterpoints' editorial policy is based on these principles and the ability of scholars to break new ground, to open new conversations, to go where educators have never gone before.

For additional information about this series or for the submission of manuscripts, please contact:
        Joe L. Kincheloe & Shirley R. Steinberg
        c/o Peter Lang Publishing, Inc.
        275 Seventh Avenue, 28th floor
        New York, New York 10001

To order other books in this series, please contact our Customer Service Department:
        (800) 770-LANG (within the U.S.)
        (212) 647-7706 (outside the U.S.)
        (212) 647-7707 FAX

Or browse online by series:
        www.peterlangusa.com